NUMBER TWO:

The Texas Engineering Experiment Station Monograph Series

ANDREW R. MCFARLAND, *Editor-in-chief*

Texas A&M University

Biomass Energy

Biomass Energy
A Monograph

Edited by Edward A. Hiler
and Bill A. Stout

PUBLISHED FOR THE
TEXAS ENGINEERING EXPERIMENT STATION,
THE TEXAS A&M UNIVERSITY SYSTEM,
BY
TEXAS A&M UNIVERSITY PRESS
COLLEGE STATION

Library of Congress Cataloging in Publication Data
Main entry under title:

Biomass energy.

(Texas Engineering Experiment Station monograph series ; no. 2)
Includes bibliographies and index.
1. Biomass energy. I. Hiler, Edward A. (Edward Allan) II. Texas Engineering Experiment Station. III. Series.
TP360.B5875 1985 662′.8 84-40559
ISBN 0-89096-231-6

Manufactured in the United States of America

FIRST EDITION

Contents

Preface

Biomass-derived fuels provide a renewable option for meeting a portion of our future energy needs. The principal goal of this monograph is to present a review of the status of biomass as an alternative energy source, with particular emphasis on the energy research programs of the Texas A&M University System.

The overall TAMU System biomass energy program represents a broad commitment to excellence in the area of alternative energy from biomass. Organizational parts of the TAMU System providing support to this program include the Texas Agricultural Experiment Station, Texas Engineering Experiment Station, Center for Energy and Mineral Resources of Texas A&M University, Texas Agricultural Extension Service, Tarleton State University, and Colleges of Agriculture and Engineering of Texas A&M University. External agencies providing support include the Texas Energy and Natural Resources Advisory Council, the U.S. Department of Energy, the U.S. Department of Agriculture, and the National Science Foundation.

Most of the TAMU System research presented herein was accomplished by multidisciplinary teams. Specific groups emphasized thermochemical conversion (combustion, gasification, pyrolysis), biological conversion (anaerobic digestion, fermentation), and plant oil extraction (physical expelling, solvent extraction). Faculty involved represent the TAMU Departments of Agricultural Engineering, Chemical Engineering, Mechanical Engineering, Forest Science, Agricultural Economics, Soil and Crop Sciences, and Chemistry; the Food Protein Research and Development Center of the Texas Engineering Experiment Station; and the Department of Agriculture at Tarleton State University.

As a result of these joint research efforts, each chapter in this monograph has a different set of authors, and most chapters have joint authorship. We believe that the team approach has greatly enhanced overall research accomplishment.

It is apparent that a wealth of technical knowledge on alternative fuels from biomass has been developed during the past decade. The extent to which that knowledge is utilized will depend largely on the price and continuing availability of petroleum-based fuels.

It is our hope that this monograph will be of value in increasing or updating the reader's knowledge of biomass energy technology.

EDWARD A. HILER

College Station, Texas
July 23, 1984

Biomass Energy

CHAPTER 1

Introduction

BILL A. STOUT

Biomass is defined as all organic matter except fossil fuels: that is, all crop and forest materials, animal products, microbial cell mass, residues, and by-products that are renewable on a year-to-year basis. Biomass serves as food, feed, fiber, bedding, structural material, soil organic matter, and fuel. The OPEC oil embargo of 1973, subsequent energy price increases, and uncertain fuel supplies have focused attention on the fundamental importance of energy in our economy and our lives. Extensive research and development programs have been undertaken during the past decade to improve energy use efficiency and develop alternative energy forms.

Overview of the U.S. Energy Situation

Table 1.1 shows ten-year trends in energy production, consumption, imports, and exports for the United States. U.S. energy consumption reached a high of 83.2 EJ (78.9 Quads)[1] in 1979. Sector use was as follows (USDOE, 1982, 1983):

Residential/commercial	19 percent
Transportation	25 percent
Industrial	26 percent
Electrical generation	30 percent
	100 percent

The entire food system—including production, processing, transportation, marketing, and final preparation—consumes about 16 percent of the nation's total energy use, with about 3 percent used in production agriculture.

Domestic oil and gas, plus imported oil, constitute nearly three-fourths of the U.S. energy supply. Oil imports are of concern because of their cost and uncertain availability. Total oil imports reached a peak

[1] Exajoule (EJ) = 10^{18} J; Quad = Quadrillion Btu or 10^{15} Btu.

TABLE 1.1 U.S. Energy Summary (exajoules)

Year	Production	Consumption	Imports	Exports
1973	65.9	78.7	15.5	2.2
1974	64.6	76.8	15.2	2.3
1975	63.4	74.6	14.9	2.5
1976	63.4	78.6	17.7	2.3
1977	63.6	80.5	21.2	2.2
1978	64.6	82.5	20.4	2.0
1979	67.4	83.2	20.7	3.1
1980	69.1	80.1	16.9	3.9
1981	68.7	77.9	14.7	4.5
1982	67.2	74.7	12.6	4.9

Source: U.S. Department of Energy (1982, 1983).

of 21.2 EJ in 1977 (nine million barrels per day); they have declined significantly since then, as conservation measures have taken effect and business has slowed as a result of recession.

Prior to the OPEC oil embargo of 1973, the world price of oil was about \$3/barrel (\$19/tonne). Figure 1.1 depicts the dramatic increase in world oil price that has occurred since 1973—a tenfold increase in a decade. While some oil price reduction has taken place recently, economists foresee increasing energy prices in the future. The era of cheap energy has passed. Thus, all energy consumers will have to cope with higher energy prices and uncertain supplies in the future.

Coping with Uncertainty

One of the most difficult dimensions of energy policy is the uncertainty of future energy prices and supplies. We must make appropriate investments in new energy production and at the same time adjust consumption to the changing patterns of energy availability in a timely manner to avoid very serious economic and social disruptions. Uncertainty makes timely investment and economic adjustment very difficult. We simply do not know, for example, what technology will develop or what will influence the supply or price of energy in the future. Public and private planning and policy must be made in the context of this uncertainty (Shaffer, 1979).

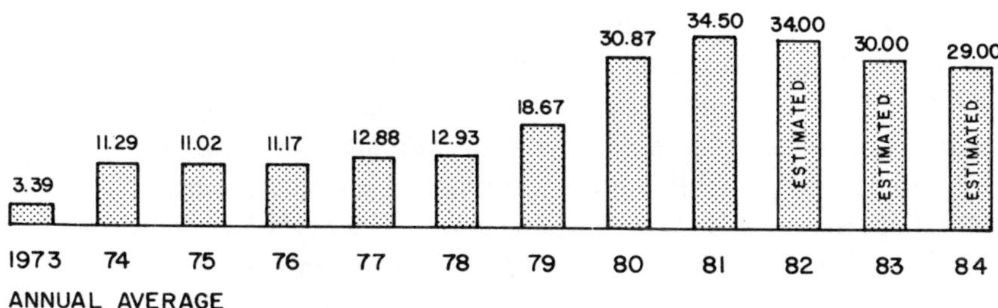

FIG. 1.1. Average OPEC crude oil official sales price, $/barrel (from USDOE, 1982, 1983).

We have had sufficient experience with temporary energy shortages to begin to understand such disruptive events as these:

—the oil embargo of 1973, which resulted in lines at gas stations, no-gas signs, and closed stations;
—the natural gas crises of 1975;
—the coal strike of 1976;
—the Iranian oil cutoff of 1978 and 1979;
—gasoline shortages in some areas of the country in 1979;
—uncertain diesel fuel supplies in the spring of 1979.

These "mini-crises" are likely to become a permanent part of our future world as the stress on finite world oil resources increases.

The Need for Alternatives

No single fuel source is likely to meet future energy needs. Rather, a great diversity of options is desirable.

Fortunately, the United States has large reserves of petroleum and natural gas, and even larger reserves of coal. Development of these domestic energy reserves, in order to reduce dependence on imported oil, has been a matter of high national priority.

Thousands of engineers and scientists are working on energy research and programs to develop new and improved technology that will effectively utilize domestic oil, gas, and coal, as well as nuclear power.

Since fossil fuel supplies are finite, energy consumption at present levels must eventually depend on renewable or inexhaustible sources, such as solar energy or nuclear fusion. Numerous discussions and debates have taken place regarding the assumptions, strategies, policies, and timetable for shifting from finite to nearly infinite energy forms.

Agriculture is an energy conversion process: through photosynthesis, crops convert solar energy to biomass, thus providing food, feed, fiber, and possibly fuels. Efforts to assess the potential of biomass fuels include the U.S. Department of Energy's Domestic Policy Review of Solar Energy (1979), the Office of Technology Assessment's study entitled "Energy from Biological Processes" (1980), and the projections for the next two decades made by the DOE Energy Research Advisory Board's Biomass Panel (Pimentel et al., 1983). The ERAB Biomass Panel estimates that there will be 700 million tons of dry biomass potentially available for fuel in the year 2000. If the available biomass is used solely as a fuel for direct combustion, a total net-heat-energy of about 5 EJ/yr will be produced. If, however, the biomass is converted to liquid or gaseous fuels, the ERAB panel estimates that the net energy available will be between 0.8 and 4.1 EJ/yr.

Biomass Energy Conversion

Many processes or technologies exist for converting biomass to useful energy or fuel forms (Stout, 1982). Figure 1.2 presents some options for converting biomass to heat energy through direct combustion, or to liquid or gaseous fuels through thermochemical, extraction, or biological processes. All of these processes are discussed in detail in later sections.

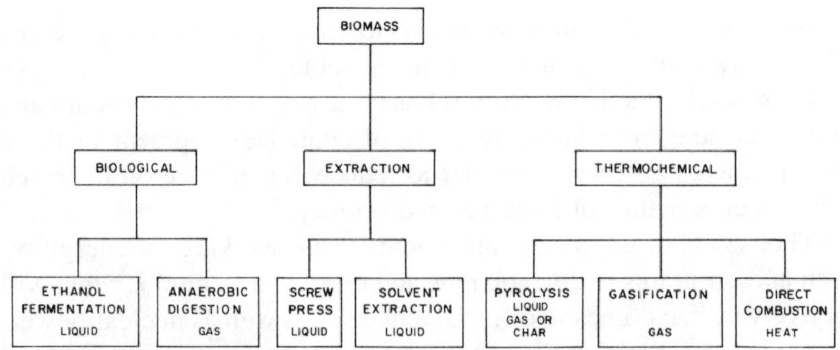

FIG. 1.2. Some options for converting biomass to heat energy or liquid or gaseous fuel (from Stout, 1984).

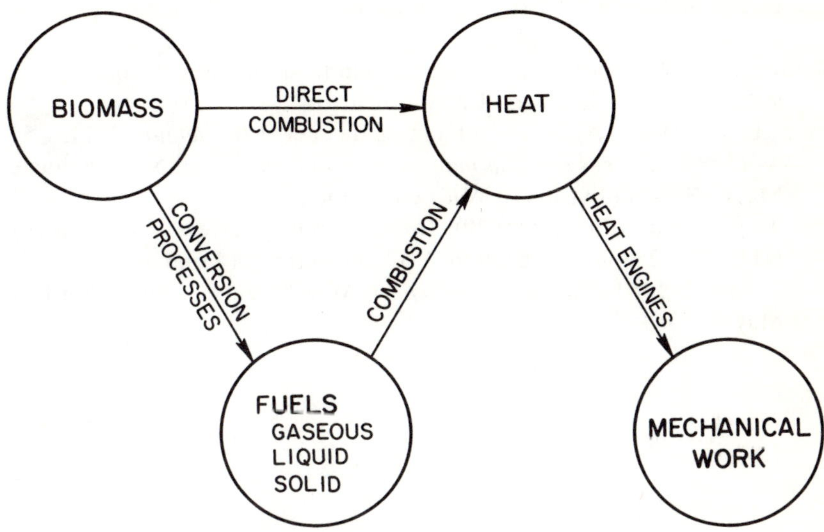

FIG. 1.3. Biomass conversion to heat or mechanical work (from Stout, 1984).

Energy is defined as the capacity to do useful work. Heat is one energy form. Energy may be converted to mechanical work by various types of heat engines.

The sugar and starch, cellulose and lignin, and other constituents of biomass may be burned directly to produce heat or mechanical work in an external combustion engine. Biomass may also be converted to liquid or gaseous fuels, which may be stored and subsequently burned to produce heat or to fuel an internal combustion engine that produces mechanical work (Figure 1.3).

Biomass energy research within the Texas A&M System encompasses most aspects of plant production, conversion of biomass to liquid or gaseous fuels, and fuel utilization; it includes economic considerations as well as integrated systems. The following chapters present a review of research on biomass as an alternative energy source.

References

Office of Technology Assessment (1980). *Energy from biological processes*. 2 vols. Washington, D.C., OTA.

Pimentel, D., et al. (1983). Biomass energy review (based on report of DOE

Energy Research Advisory Board, Biomass Panel). *Solar Energy* 30 (1): 1–31.

Shaffer, J. (1979). Energy, transportation and food: Interdependent policy issues. Rochester Seminars. Unpublished.

Stout, B. A. (1982). Agricultural biomass for fuel. *Experientia* 38: 145–51.

——— (1984). *Energy use and management in agriculture*. North Scituate, Mass.: Breton (division of Wadsworth, Inc.).

U.S. Department of Energy (1979). Domestic policy review of solar energy. NTIS TID-28834. Washington, D.C.: Government Printing Office.

——— (1982, 1983). *Energy users report*. May 13, June 3, 1982; April 31, May 5, 1983.

CHAPTER 2

Thermochemical Conversion for Energy and Fuel

WAYNE A. LEPORI and ED J. SOLTES

Current sources of energy are obtained primarily from fossil fuels. These fuels are stable chemical compounds which can release or absorb thermal energy, as atoms and/or molecules of compounds combine to produce other chemical compounds. Using heat and/or catalysts to create a change in chemical form is termed thermochemical conversion. Today, the use of thermochemical processes is the most important method of converting the chemical energy of fuels to more desirable energy forms.

Fossil fuels in the form of coal, oil, and natural gas are the predominant sources for such thermochemical conversion. However, this has not always been true. Not only is biomass believed to have been the original source of fossil fuels, but controlling the burning of wood for heating and cooking was probably one of the first energy conversion processes used for the benefit of humankind. The recent realization of the limitations of fossil fuel supplies has created renewed interest in using biomass energy forms through thermochemical processes.

Present equipment and fundamental thermochemical conversion techniques have developed around the fossil fuels. Society has had comparatively little experience in the thermochemical conversion of biomass—with the exception of wood. The erroneous assumption that all biomass materials are the same, and will react the same as wood or fossil fuels in thermochemical conversion processes, has led to failures in the use of certain other biomass materials as fuel.

Biomass Fuels for Thermochemical Processes

Biomass, like the fossil fuels, contains high percentages of carbon and hydrogen. Biomass fuels, in contrast to fossil fuels, are also high in oxygen. Ultimate analysis, which determines the mass fraction of carbon (C), hydrogen (H), oxygen (O), sulfur (S), nitrogen (N), and ash, has been used in evaluating fuels. Ultimate analyses of several biomass and fossil fuel types can be compared in Table 2.1

TABLE 2.1 Ultimate Analysis Data for Selected Solid and Biomass Fuels

Material	Elements (% dry weight)						Gross heating value (kJ/kg)	References
	C	H	N	S	O	Ash		
Pittsburgh seam coal	75.5	5.0	1.2	3.1	4.9	10.3	31,721	Reed (1981)
West Kentucky No. 11 coal	74.4	5.1	1.5	3.8	7.9	7.3	31,280	Reed (1981)
Utah coal	77.9	6.0	1.5	0.6	9.9	4.1	32,930	Reed (1981)
Wyoming Elkol coal	71.5	5.3	1.2	0.9	16.9	4.2	29,537	Reed (1981)
Lignite	64.0	4.2	0.9	1.3	19.2	10.4	24,894	Reed (1981)
Charcoal	80.3	3.1	0.2	0.0	11.3	3.4	57,390	Reed (1981)
Douglas fir	52.3	6.3	0.1	0.0	40.5	0.8	21,031	Reed (1981)
Douglas fir bark	56.2	5.9	0.0	0.0	36.7	1.2	22,077	Reed (1981)
Pine bark	52.3	5.8	0.2	0.0	38.8	2.9	20,404	Reed (1981)
Western hemlock	50.4	5.8	0.1	0.1	41.4	2.2	20,032	Reed (1981)
Redwood	53.5	5.9	0.1	0.0	40.3	0.2	21,008	Reed (1981)
Beech	51.6	6.3	0.0	0.0	41.5	0.6	20,357	Reed (1981)

Hickory	49.7	6.5	0.0	0.0	43.1	0.7	20,148	Reed (1981)
Maple	50.6	6.0	0.3	0.0	41.7	1.4	19,939	Reed (1981)
Poplar	51.6	6.3	0.0	0.0	41.5	0.6	20.729	Reed (1981)
Rice hulls	38.5	5.7	0.5	0.0	39.8	15.5	15,361	Reed (1981)
Rice straw	39.2	5.1	0.6	0.1	35.8	19.2	15,198	Reed (1981)
Sawdust pellets	47.2	6.5	0.0	0.0	45.4	1.0	20,483	Reed (1981)
Paper	43.4	5.8	0.3	0.2	44.3	6.0	17,597	Reed (1981)
Redwood wastewood	53.4	6.0	0.1	39.9	0.1	0.6	21,294	Reed (1981)
Alabama oak wastewood	49.5	5.7	0.2	0.0	41.3	3.3	19,196	Reed (1981)
Animal waste	42.7	5.5	2.4	0.3	31.3	17.8	17,150	Reed (1981)
Municipal solid waste	47.6	6.0	1.2	0.3	32.9	12.0	19,860	Reed (1981)
Cotton gin trash	42.0	5.4	1.4	<0.5	35.0	14.5	15,500	LePori et al. (1981a)
Sorghum stalks	40.0	5.2	1.4	0.2	40.7	12.5	15,400	LePori et al. (1981a)
Feedlot manure (fresh)	45.4	5.4	1.0	0.3	31.0	15.9	17,361	Sweeten et al. (1983)
Feedlot manure (aged, composted)	33.0	4.9	0.7	0.8	17.5	41.6	15,124	Sweeten et al. (1983)
Corncobs	46.2	7.6	1.2	0.3	42.3	2.4	26,328	Lee (1981)
Rice hulls	41.3	8.4	1.0	0.02	33.0	18.3	16,513	Lee (1981)
Sawdust	49.7	6.2	0.7	0.17	42.6	0.7	19,970	Datin (1983)

It is evident from ultimate analysis that differences exist between fossil fuels and biomass materials, and between the various biomass materials. Biomass materials generally contain lower percentages of carbon and higher percentages of oxygen, resulting in a lower heating value for biomass than for fossil fuels. Ratios of hydrogen-to-carbon and oxygen-to-carbon are higher for biomass fuels than for fossil fuels. These ratios provide insight into simple reaction processes such as dehydration, deoxygenation, and hydrogenation. They also show that more biomass than fossil fuel must be processed to obtain an equal quantity of energy.

Biomass fuels are much more reactive than most coal fuels and are from 70 to 90 percent volatiles (Reed, 1981). Their ratios of volatile carbons to fixed carbons range from 3–9:1, while coals range from 0.1–1:1 (Sarkanen, Tillman, and Jahn, 1982). These differences must be considered in thermal conversion techniques and equipment.

Combustion equipment must be designed to control the release of the volatiles; otherwise, they can escape without being completely utilized in the combustion process. A longer reaction time for fixed carbon is essential for maintaining temperature and igniting the volatiles as they are released during the combustion process. In contrast, high volatility increases the ease with which a fuel may be gasified.

In addition to energy content, there are several other important considerations in thermochemical conversion. These include fuel preparation (such as particle-size reduction), ash-softening temperature, and hazardous impurities. Ash-softening temperature is the temperature where the ash becomes plastic; a fuel with a low ash-softening temperature can create fused masses called clinkers, which interfere with the thermochemical process. Because of the wide variation in biomass materials, knowledge of some fundamental properties is lacking, even though such knowledge is critical in determining how such materials will function as fuels with present thermal conversion technology and equipment.

Thermochemical Processes

All complex solid organic material undergoes decomposition upon heating, and forms gaseous, liquid, and solid fractions. The relative quantity of each fraction can be varied by controlling various parameters during the thermal decomposition process.

The thermal techniques generally used in controlled conversion processes are (1) direct combustion, (2) gasification, and (3) pyrolysis. These terms have various definitions, but they will be used here to identify the type of end-product desired. *Direct combustion* is a complete oxidation process, where liberation of heat is the primary objective. *Gasification* is a partial oxidation process that results primarily in combustible gases. *Pyrolysis* is a nonoxidative thermal process that results in gases, liquids, and char; all three products can be used as fuels or chemical feedstocks, but the objective of pyrolysis is generally to maximize either the liquid or the solid fraction.

Numerous parameters can be controlled to differentiate between processes and to obtain the desired end product. These include heating rate, final temperature, residence time at temperature, presence or absence of air or oxygen, fuel particle size, and fuel moisture content. Fuel-to-air ratio and operating temperature are perhaps the two most critical parameters to control in biomass conversion processes.

Direct Combustion Fundamentals

The objective of direct combustion of fuels is to obtain heat, which is then used directly or through a secondary process. Heat is liberated by the chemical changes that occur when oxygen reacts with carbon and hydrogen in a fuel, forming carbon dioxide and water. Sulfur is generally ignored in the direct combustion of biomass fuels because of the low sulfur content of biomass materials. The chain of reactions and heat released during direct combustion of hydrocarbons is illustrated by the equations below (Culp, 1979):

$$2C + O_2 \rightarrow 2CO + 2Q_{C\rightarrow CO}$$

$$2CO + O_2 \rightarrow 2CO_2 + 2Q_{CO\text{-}CO_2}$$

$$2H_2 + O_2 \rightarrow 2H_2O + 2Q_{H_2\text{-}H_2O}$$

$$Q_{C\rightarrow CO} = 110{,}380 \text{ kJ/(kg·mol C)}$$

$$Q_{CO\rightarrow CO_2} = 283{,}180 \text{ kJ/(kg·mol CO)}$$

$$Q_{H_2\rightarrow H_2O} = 286{,}470 \text{ kJ/(kg·mol } H_2)$$

If combustion temperature is very high, some heat-of-the-process would be absorbed in processes that break down some of the compounds, such as carbon dioxide. This will generally not occur in biomass conversion processes, but heat is commonly absorbed to vaporize moisture in the material.

Three factors are required for the complete combustion of a fuel: (1) heat to initiate the reaction, (2) mixing of oxygen or air with fuel, and (3) sufficient time for the completion of reactions. Perfect conditions seldom exist, and in addition to the products of complete combustion, some incomplete-combustion products result. These include unburned fuel, carbon monoxide, and nitrogen compounds.

To minimize many of the incomplete combustion products, excess oxygen is supplied. The theoretical quantity of air necessary to provide sufficient oxygen to completely combine with a quantity of fuel is termed *stoichiometric air*. Any air beyond this quantity is termed *excess air*. Selection of air-to-fuel ratios is important because too much excess air can have detrimental effects on direct combustion:

(1) cooling and slowing chemical reaction rates;
(2) increasing exit-gas velocities, thus carrying partially burned particles out of the combustion chamber; and
(3) increasing power requirements to move air through the combustion chamber.

Actual combustion of organic residues is not a single process but a combination of processes occurring simultaneously. Three overlapping phases can be identified: (1) evaporation of moisture, (2) volatilization and burning of volatiles, and (3) burning of fixed carbon. Evaporation of moisture is the initial phase of combustion. Next, the dry matter absorbs heat, driving off volatile gases, which rise and burn. Finally, the fixed carbon burns in the last phase. The direct-combustion process is illustrated in Figure 2.1.

High moisture content can reduce the combustion efficiency of

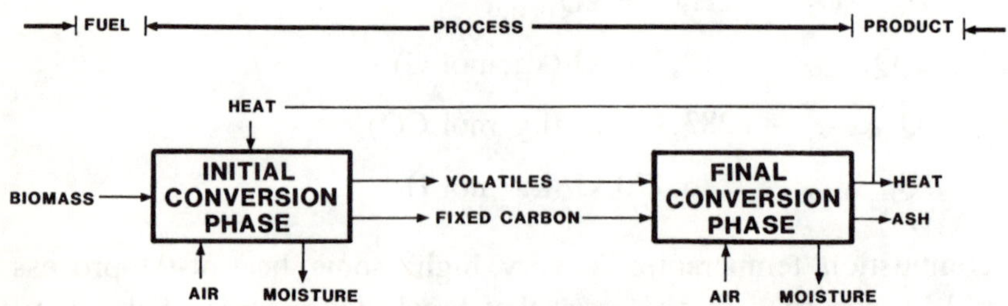

FIG. 2.1. Schematic of the direct combustion process.

biomass fuels. For example, a kilogram of fuel with a 50 percent moisture content contains 500 grams of moisture to be evaporated. Changing this quantity of moisture from a liquid to a vapor requires about 525 kilojoules of heat, which represents about 14 percent of the available energy in the 500-gram dry-matter portion of the fuel. Thus, increasing moisture content not only reduces the quantity of dry fuel available per pound of material but also uses a significant portion of the energy to remove the moisture. Evaporation of moisture absorbs that energy which is needed to maintain adequate temperature and to sustain the combustion process.

Gasification and Pyrolysis Fundamentals

Pyrolysis is a chemical change brought about by the action of heat and is also an important stage in all gasification and combustion processes (Reed, 1981). The products of pyrolysis are gases, liquids, and char; the relative quantity of each depends on fuel properties, rate of heating, and final temperature attained. Pyrolysis normally proceeds at temperatures near 600°C, while gasification processes occur in the temperature range of 800 to 1100°C (Probstein and Hicks, 1982). Gasification processes generally use reactants such as oxygen or steam to increase gas yields while consuming char.

Gasification and pyrolysis involve several chemical reactions, which occur simultaneously at varying rates. In systems where solid fuels are gasified in the presence of substoichiometric air, the principal gaseous products include hydrogen, carbon monoxide, carbon dioxide, nitrogen, methane, and small amounts of other hydrocarbons. The products are formed by a combination of the following chemical reactions:

$$C + O_2 \rightarrow CO_2$$

$$2H_2 + O_2 \rightarrow 2H_2O$$

$$C + H_2O \rightarrow CO + H_2$$

$$C + CO_2 \rightarrow 2CO$$

$$CO + H_2O \rightarrow CO_2 + H_2$$

$$C + 2H_2 \rightarrow CH_4$$

$$C_nH_m + \text{Heat} \rightarrow CH_4 + C_2H_4 + C_2H_6 + C_3H_6 + C_3H_8 + \ldots$$

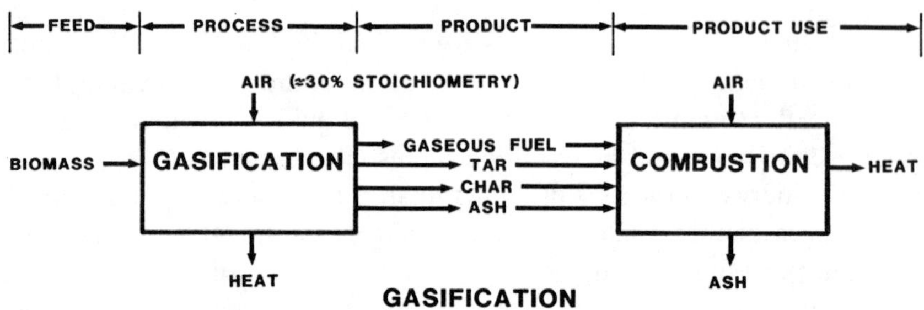

FIG. 2.2. Schematic showing the gasification process.

In air gasification, nitrogen is the primary component of the gas and has a diluting effect on the combustible components. The heating value of this low-calorific-value (LCV) gas is generally about 10 to 20 percent that of natural gas. In addition to gaseous fuels, condensible liquids (termed *tars*) and unreacted carbon (termed *char*) will be produced in an air gasification process (Figure 2.2). For most end uses of the combustible gas, it is desirable to minimize tar and char formation and maximize gaseous fuel production. This generally represents the distinction between gasification and pyrolysis, which is discussed in the next section.

The relative amount of each component of LCV gas is a function of many variables. The three most important variables are the reaction temperature, the amount of air or oxygen present, and the time at which the products remain at reaction temperature. Numerous methods and types of equipment have been developed to control these parameters, and each method has different characteristics with respect to controlling the three primary variables.

An equivalence ratio is often used to express the quantity of air relative to fuel in a gasification process. It is defined as the oxidant-to-fuel weight ratio divided by the stoichiometric ratio. Gasification systems generally operate in the range of 0.20 to 0.40 for maximum production of combustible gas.

Combustion Technologies

In agriculture and forestry, the objective of direct combustion is to produce heat for space heating, crop drying, steam production, or a combination of these (Smith, Riley, and Schaufler, 1983; Henry,

Keener, and Anderson, 1983; Claar, Buchele, and Morley, 1981; LePori et al., 1981c). It is currently the most widely used process for converting biomass to usable energy. Numerous designs and variations of equipment exist for the direct combustion of biomass fuels (Sarkanen, Tilman, and Jahn, 1982). Each design has its unique characteristics, but most were developed for burning fossil fuels, primarily coal. These systems are generally classified as pile burners, semi-pile burners, suspension burners, and fluidized-bed combustors. Schweiger (1980) has discussed their design and operating principles.

Several systems have recently been designed specifically to control reactions and efficiently burn biomass materials. Claar, Buchele, and Morley (1981) have designed and tested a concentric-vortex cell furnace that provides staged combustion by controlling air injection at several points within the unit. They have successfully substituted corncobs for LPG (liquefied petroleum gas) as fuel in a crop dryer. Henry, Keener, and Anderson (1983) also used corncobs in a fluidized bed to produce heat. Maximum temperatures were limited to 760°C to prevent ash and bed-particle agglomeration.

Several other pile-burning, incinerator-type furnaces for burning crop residues have been reported. Burners for large round hay bales have also recently been offered commercially, but little performance data is available on these. One bale burner, a two-stage unit that actually gasifies large round bales in one chamber and burns the gases in another, has been used for supplying heat in grain drying (Claar, Buchele, and Morley, 1981; Page, 1979). Smith, Riley, and Schaufler (1983) have developed efficient wood-burning furnaces for space heating.

Wood has been the predominant biomass fuel used in direct combustion systems. As other biomass fuels are used, it becomes evident that different biomass materials react differently in the same equipment. Some, particularly those with high ash content, can create slag or ash scale buildup on system components (Lang, 1980).

Several studies have described the fundamental mechanism of slag or ash deposition on system components (Wall, Graves, and Roberts, 1975; Ballantyne et al., 1980). These studies show that complex chemical reactions can form eutectic mixtures which have melting points lower than the ash-softening temperature. Eutectics composed of elements found in the ash of many biomass fuels were shown to have melting points below 612°C. Combustion normally occurs at temperatures which exceed the melting point of eutectic compounds; therefore, slagging and ash deposition can be expected from burning biomass fuels.

Elements in the ash that can create eutectic mixtures in an oxidizing atmosphere include potassium, calcium, magnesium, silica, sodium, iron, and aluminum. Attig and Duzy (1969) have developed slagging and fouling indexes for coals having differing ash chemical composition. The term *slagging* was used to identify deposits that formed on furnace walls and other surfaces exposed to predominantly radiant heat; *fouling* was used to identify deposits that formed on surfaces not exposed to radiant heat from the furnace. The indexes were based on ratios of basic to acid oxide constituents. The base-to-acid ratio was multiplied by the percentage of sodium oxide to obtain a fouling index and of sulfur to obtain a slagging index. Fuels with fouling index values less than 0.2 were considered low fouling; those with values greater than 0.5, high fouling. Slagging indexes of less than 0.6 indicated low-slagging fuels; values greater than 2.0, high-slagging fuels.

LePori et al. (1981a) have identified slagging and fouling as serious problems in the direct combustion of cotton gin trash and sorghum stalks. A conventional atmospheric fluidized bed, where bed temperatures were maintained near 760°C, was used in their study. The chemical composition of ash and fouling deposits included high concentrations of silica, potassium, calcium, phosphorus, aluminum, and sulfur, as shown in Table 2.2. Applying the coal slagging and fouling index from Attig and Duzy (1969) to these data shows cotton gin trash to be a severe-slagging and medium-fouling fuel, sorghum a low-slagging and -fouling fuel, and fresh manure a high-fouling and severe-slagging fuel. Combustion tests showed that all three were actually high-slagging and -fouling fuels, with sorghum stalks being more severe than cotton gin trash and manure more severe than sorghum stalks.

Ash deposition from biomass fuels containing certain chemicals can also create corrosion and erosion of metals. LePori et al. (1981b) reported that the same chemicals that created slagging and fouling problems also caused erosion and corrosion of metal coupons. They showed that low carbon steel 304, stainless steel, Incoloy 800, and Incoloy 825 suffered severe corrosion damage when exposed to hot exhaust gases from the combustion of cotton gin trash and sorghum

TABLE 2.2 Analysis of Ash and Fouling Deposits from Fluidized-Bed Combustion of Biomass Fuels

Fuel	Compounds (%)								
	SiO_2	K_2O	CaO	MgO	Na_2O	P_2O_5	Fe_2O_3	Al_2O_3	SO_3
Cotton gin trash									
Ash	41.8	10.5	10.8	3.3	0.6	2.6	0.7	3.1	5.9
Deposits	15.4	34.3	20.2	8.6	1.5	4.6	0.6	1.5	9.1
Sorghum stalks									
Ash	73.2	8.4	5.0	1.5	0.4	1.1	1.0	5.1	0.5
Deposits	5.4	33.3	6.5	1.5	0.3	0.8	1.9	4.0	2.6
Feedlot manure									
Ash (fresh)	53.5	6.4	13.9	3.7	2.0	3.0	1.7	7.8	2.8
(aged, composted)	59.5	4.2	7.2	2.3	1.4	4.7	1.3	9.8	3.7
Deposits									
(fresh)	65.9	1.9	3.2	0.9	0.7	3.1	1.0	6.8	1.8
(aged, composted)	67.9	2.2	3.3	1.1	1.8	2.9	1.4	7.1	2.3

stalks. Analyses showed that corrosive reactions occurred between chemicals in particles of the exhaust gases and the elements in the metal. These corrosive reactions occurred on uncooled samples at gas temperatures near 650°C.

Summary: Direct Combustion

Direct combustion is at present the most widely used technique for converting biomass materials to energy; however, successful systems have primarily used wood or forest products as fuel. Recent experience with other high-ash biomass fuels shows that slagging and fouling can be a severe limitation in using direct combustion conversion equipment. Caution is necessary in extrapolating the feasibility of equipment designed for one biomass fuel to the use of another. In addition to ultimate analyses, chemicals in the ash including basic and acidic constituents are important considerations in evaluating the suitability of biomass fuels for direct combustion.

Gasification Technologies

Gasifiers can be classified as moving-bed, fluidized-bed, rotary-kiln, and multiple-hearth types. These classifications, based on the method by which fuel is moved through the gasifier, have been discussed by Beck (1979) and Reed (1981). Moving-bed gasifiers can be further identified according to the direction in which air moves through the bed—downdraft, updraft, or crossdraft—and there are many variations of these, such as the channel-flow gasifier (Richey, Barrett, and Klutz, 1983).

Moving-bed gasifiers require fuels that allow free and uniform passage of gas through the fuel bed. Such fuels must be relatively uniform in particle size so that the gases do not form channels; therefore, many biomass fuels require special preparation, such as pelleting, before they can be used. High-temperature zones exist in the regions where the air is introduced, and ash-fusion temperatures can be exceeded for some biomass fuels (Williams and Goss, 1979), causing fouling of grates.

Many research studies have focused on moving-bed gasifiers, and certain types have been used commercially with success (Reed, 1981). Wood is the predominant fuel for these gasifiers, but other fuels have been evaluated. Much of the development of gasifier technology occurred in Europe during World War II, when supplies of conventional fuels were limited.

Williams and Goss (1979) used Swedish design concepts to develop a downdraft gasifier. They have evaluated many biomass fuels, reported on preparation techniques needed for each fuel, and also identified fuels that could not be gasified because of slagging. Richey, Barrett, and Klutz (1983) have developed a variation of the downdraft gasifier called a channel-flow gasifier. This design would appear to provide more opportunity for temperature control in the hot zone and could possibly handle some of the so-called slagging fuels.

Several two-stage gasifier/combustors have recently been introduced experimentally and commercially (Payne, Ross, and Walker, 1979; Payne et al., 1981). These use the updraft gasification technique but provide a second chamber very close to the first to burn the gas and produce heat. This method avoids condensation of tars and the resulting consequences; however, it limits end use of the gases. It, too, has been used primarily with wood.

Fluidized-Bed Gasification

Fluidized-bed technology offers several unique and versatile characteristics for biomass gasification. The turbulent, fluidized state of inert particles in the bed creates a near-isothermal zone and enables accurate control of reaction temperatures. Thermal energy stored in the large mass of inert particles is rapidly transferred to solid fuel at stable temperatures. Violent agitation of solids provides efficient conversion reactions and allows introduction of fuels having wide variations in composition and particle size.

A broad range of operating conditions can be obtained in a fluidized-bed unit. Parameters that can be easily varied include air velocity, operating pressure, type of bed material, and size of bed material. The ability of fluidized beds to accept many types of biomass materials with little fuel preparation is one of their greatest advantages.

Fuel injection into fluidized beds is achieved by various methods, which can influence residence time in the isothermal bed zone. Most systems inject feed near the top of the bed (Raman et al., 1981; Beck, 1979), while others inject feed near the base (LePori et al., 1981b; Moreno, 1982). Bottom-feed systems in fluidized beds are more analogous to downdraft moving-bed systems, which characteristically produce less tar. Tar formation is generally undesirable in gasification, because tars are sticky when condensed; they can create problems with gas cleanup devices or clog burners. Tars are decomposed into other hydrocarbons as they move through the isothermal zone in a fluidized-bed unit.

Fluidized beds have been used in gasification studies of various biomass materials (Walawender, 1983; Beck, 1979; Reed, 1981). Some systems use steam with the air as a means of increasing the heat value of the gas; others use oxygen, but most use air as the fluidizing and gasification medium. The predominant biomass fuel studied has been forest products; however, current research is being done with other materials also.

Fluidized-bed gasification tests using fuels with varying ash levels (cotton gin trash, rice hulls, corncobs, and sorghum stalks) reveal that differences exist in gas composition and gas heating values for the different fuels when operating under the same conditions (Groves, 1979; Lee, 1981). These studies used a bench-scale 51-mm-diameter unit and required fuel feedstock to be reduced in particle size to pass through a 1-mm screen. Supplemental heat was used to offset surface heat losses. Operating bed temperatures were selected within the 700 to 900°C range, and fuel-to-air ratios ranged from 0.7 kg/kg to 1.8 kg/kg. The lower fuel-to-air ratio is similar to that which can be achieved in larger units without supplemental heat.

The effect of temperature and fuel-to-air ratios on the heating value of gas produced from the four fuels is shown in Figures 2.3 through 2.6. Over the limited temperature range evaluated, the heating value of the gas was essentially independent of temperature and primarily a function of fuel-to-air ratio. As air-to-fuel ratio decreased, heating value increased, illustrating the diluting effect of nitrogen on heating value.

The relative composition of the gas was also different for the different fuels under similar operating conditions, as shown in Figures 2.7 through 2.10. One of the primary differences was the relative quantities of CO and CO_2 produced by the different fuels. Smaller differences also occurred in the quantities of CH_4 and C_2H_4, but these small differences are magnified in terms of heating value because of the higher heating value of these gas components.

Craig (1980) has shown that similar gas composition and heating values were obtained in a 305-mm-diameter fluidized bed. However, he also showed that the range of air-to-fuel ratios was limited when no

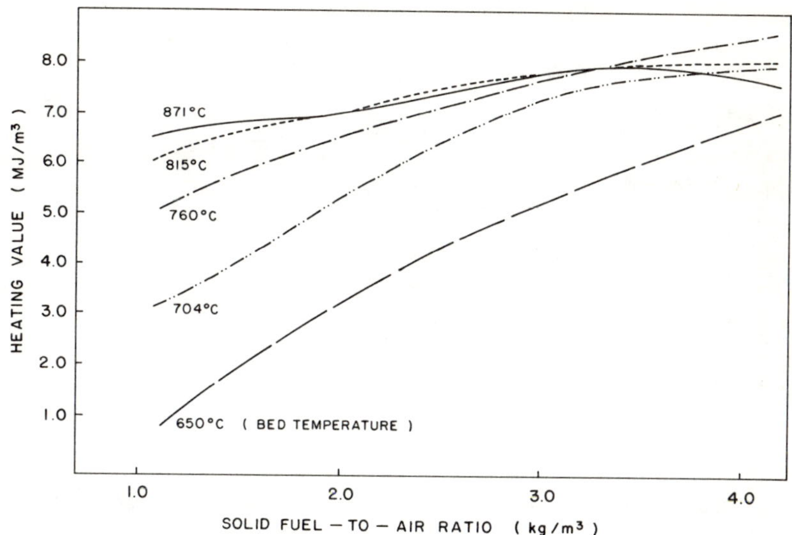

FIG. 2.3. Results of empirical correlation of the gasification of cotton gin wastes.

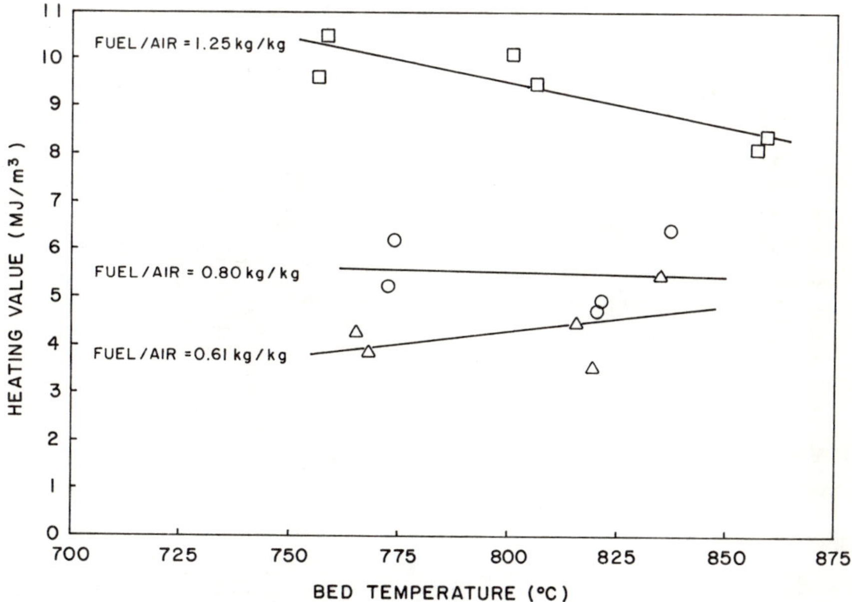

FIG. 2.4. Effect of temperature and fuel-to-air ratio on the heating value of gas produced from sorghum stalks.

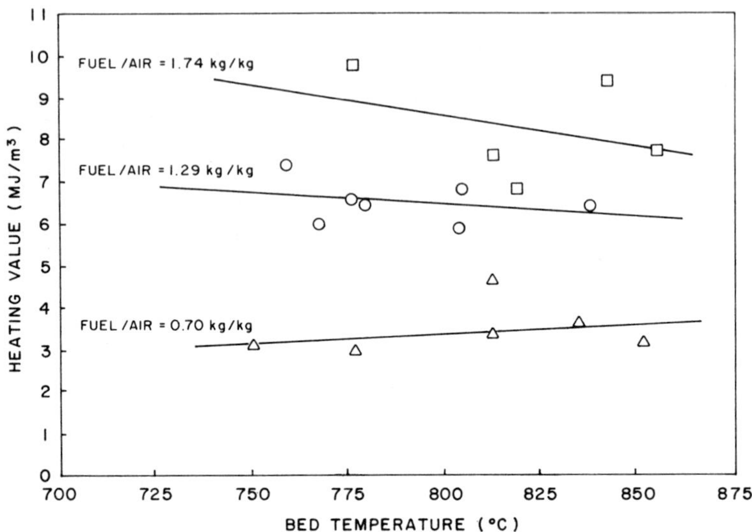

FIG. 2.5. Effect of temperature and fuel-to-air ratio on the heating value of gas produced from rice hulls.

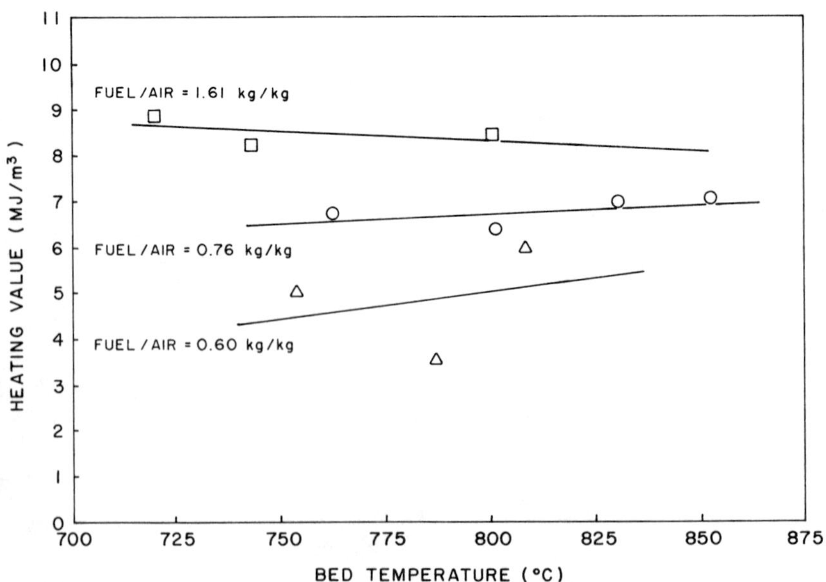

FIG. 2.6. Effect of temperature and fuel-to-air ratio on the heating value of gas produced from corncobs.

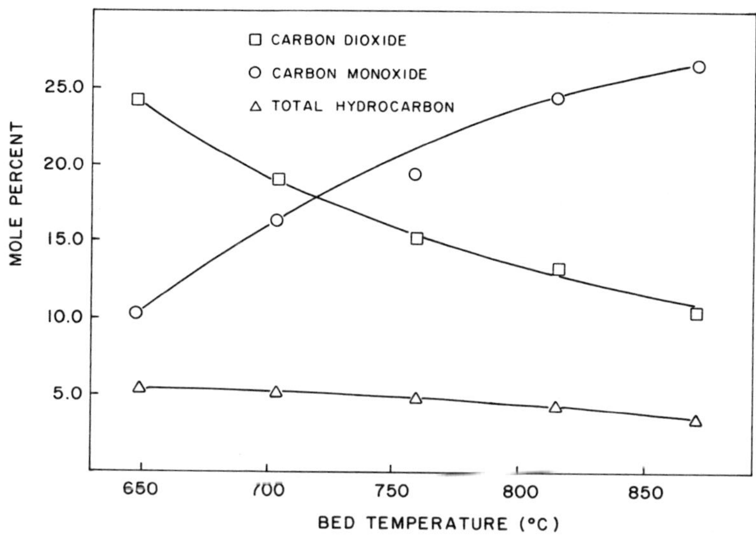

FIG. 2.7. Production of oxides of carbon and total hydrocarbons in a fluidized-bed gasifier.

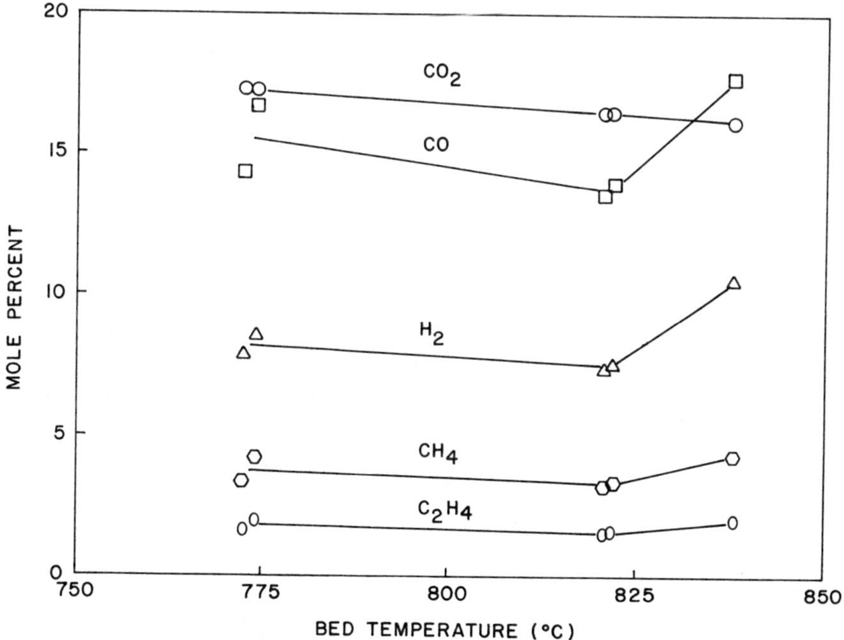

FIG. 2.8. Composition of gas (excluding nitrogen) produced from sorghum stalks at a fuel-to-air ratio of 0.8 kg/kg.

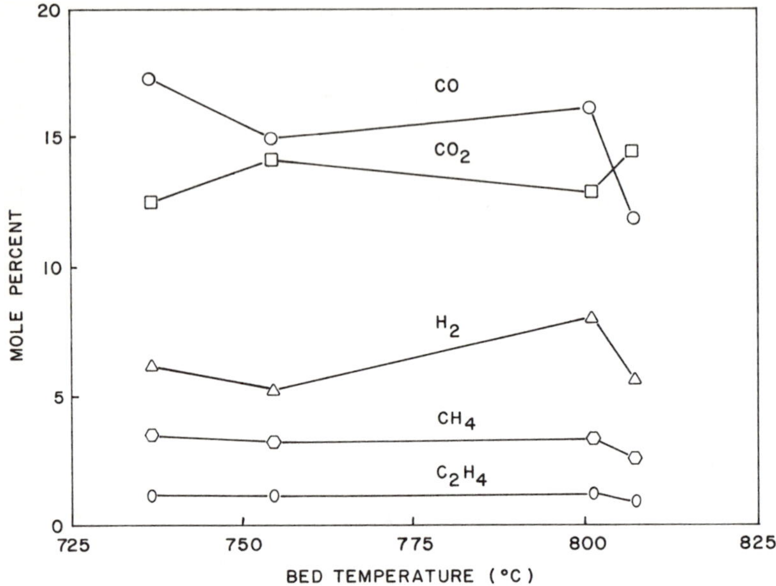

FIG. 2.9. Composition of gas (excluding nitrogen) produced from rice hulls at a fuel-to-air ratio of 0.862 kg/kg.

supplemental heat was used. With no external heat source, air-to-fuel ratio must be increased for partial combustion of fuel to generate heat for the endothermic gas reactions. The most significant result of using the fluidized-bed gasification process was the elimination of fouling and slagging.

The reducing atmosphere of the gasification process apparently inhibits formation of the low-melting-point eutectics. Fung and Graham (1980) showed that some of the chemicals, such as potassium, that cause problems in direct combustion actually serve as catalysts to increase gas production. Specifically, potassium carbonate and calcium oxide were shown to increase conversion of solid char products to combustible gas products.

These results were used to develop a fluidized-bed gasifier/boiler system to help evaluate the conversion of biomass materials to steam (Figure 2.11). This system was based on the principle of removing ash components at low temperatures in a reducing atmosphere, since removing the chemical components that create fouling and slagging

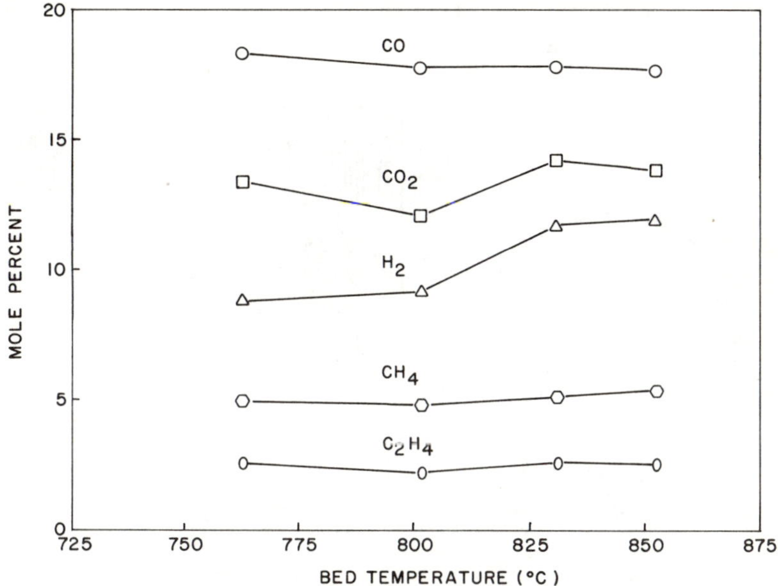

FIG. 2.10. Composition of gas (excluding nitrogen) produced from corncobs at a fuel-to-air ratio of 0.765 kg/kg.

should result in a more reliable system. Direct combustion rather than gasification could possibly be used if fuel additives were introduced to reduce slagging and fouling (Wall, Graves, and Roberts, 1975; Ballantyne et al., 1980). However, the approach of first generating a combustible gas at low temperature, removing the ash, and then burning the gas at high temperatures provides advantages in terms of more end uses for the gas.

Research Steam-Producing System

The steam-production unit is composed of five subsystems: (1) fuel feed, (2) fluidized-bed reactor, (3) cyclone particulate removal, (4) gas-mixing chamber and burner, and (5) fire-tube boiler (see Figure 2.11).

Fuel-Feed System. Biomass materials are difficult to handle, but a fuel-feed system was developed which consists of a hopper with an agitator to prevent bridging, an air lock to prevent backflow of hot gases, and two feed augers. This design permits the injection of biomass materi-

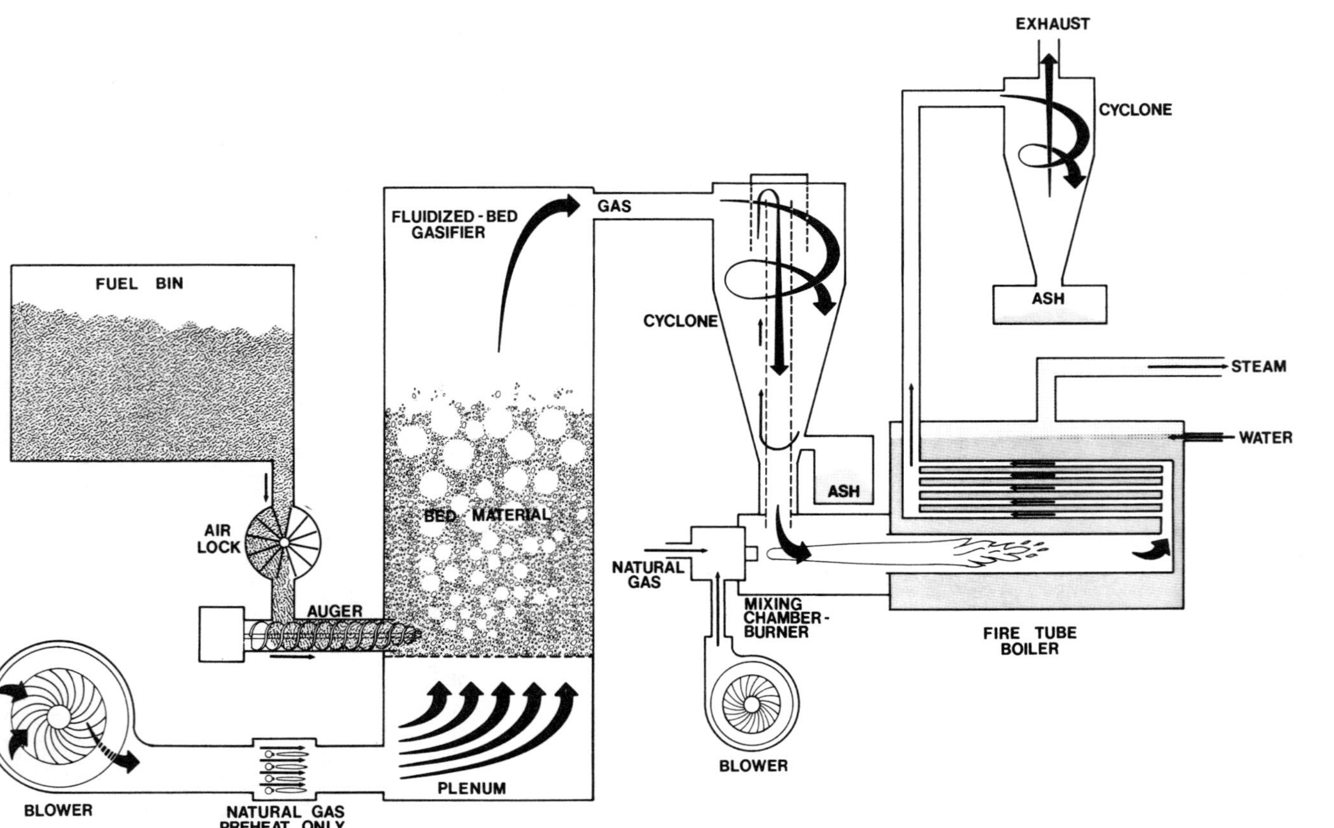

FIG. 2.11. Schematic of fluidized-bed biomass steam-producing system.

als, such as unprocessed gin trash, into the fluidized bed near the distributor plate.

The agitator designed to keep fuel from bridging in the hopper forces it into the metering auger. This first auger meters the fuel from the hopper to the air lock at a rate proportional to the auger speed. The air lock then drops the trash into a second auger, which introduces the fuel into the lower region of the fluidized bed. Air is supplied to the second auger to assist in feeding and to prevent hot gases from backflowing into the auger tube. Since the pressure in the second auger tube is higher than atmospheric pressure, the rotary air lock will always have some leakage, and the air supplied to the second auger keeps it purged of hot gases. The second auger is run at a higher speed than the first to maintain only a partial fill and to prevent plugging.

Fluidized-Bed Reactor. The 610-mm-diameter fluidized-bed unit originally used for direct combustion studies was adapted as the gasification reactor in the steam-producing system. The design consists of a positive displacement blower, natural gas burner for start-up, bubble-cap air distributor, and refractory-lined reaction chamber. The bed material is refractory sand to a depth of approximately 1 meter in an unfluidized state. Thermocouples are used to monitor temperatures at various points within the system, and pressure taps placed at various points permit pressure measurements. Fluidizing air is measured by means of a laminar flow element.

Particulate Removal. Ash and solid particles are removed from the gasifier by elutriation with the combustible gas, using two cleaning devices; the first was placed in the exit-gas stream from the fluidized bed, and the second is a cyclone in the exit-gas stream from the boiler. The first cleaning device, which removes much of the ash and the unreacted carbon particles, was originally a cyclone but was modified to provide internal transfer of combustible gas to a burner (Figure 2.11). A separate study was made to evaluate particulate removal requirements in the gasification of biomass; it is discussed below.

Gas-Mixing Chamber and Burner. Combustible gas, leaving the primary cyclone, is mixed with air and ignited in a chamber before entering the boiler. Air is introduced into a refractory-lined chamber in a

TABLE 2.3 Boiler Characteristics

Steam production	568 kg/hr
Heating surface	19.5 m^2
Firing rate gas (37.3 MJ/m^3)	42.5 m^3/hr
Boiler shell outside diameter	1.07 m
Overall length	2.92 m
Overall height	1.68 m
Furnace outside diameter	410 mm
Tube diameter	50 mm
Steam working pressure (max.)	8.62 kPa

swirling motion to mix with the combustible gas exiting the primary cyclone, and the mixture is ignited by a natural-gas pilot.

Fire-Tube Boiler. A fire-tube boiler with feed-water pump and automatic water level controls is connected to the gas-mixing chamber and burner. It was designed for natural-gas operation, but the natural-gas burner was removed and replaced by the low-energy gas unit. Boiler characteristics are shown in Table 2.3.

System Operation and Results

Operating the fluidized-bed gasifier boiler system with cotton gin trash as fuel indicated its ability to transform biomass materials into usable energy (LePori et al., 1983). Overall efficiencies of this system for several fuel input and steam production rates ranged from 42 to 56; they are shown in Table 2.4. The overall energy conversion efficiency for the system is the ratio of rate of net energy added to steam, to the fuel-energy input rate. These efficiency values do not include energy inputs of air-handling equipment, but they are corrected for the natural-gas pilot used to ignite the low-calorific-value (LCV) gas.

Energy for air handling was higher than it would be in a commercial unit because excess capacity is needed to cover the wide range of air-to-fuel ratios used in the research. The excess air in a specific test was diverted; therefore, the energy for air handling was not included.

Efficiencies were influenced by several factors, including the quality of burning the LCV gas. The efficiencies near 55 percent achieved with this experimental unit show the potential for power pro-

TABLE 2.4 Overall Energy Conversion Efficiency for steam production

Cotton Gin Trash		Steam Production Rate		Overall Efficiency	
Fuel Rate (kg/hr)	Standard Deviation	kg/hr	Standard Deviation	%	Standard Deviation
169	10	436	92	42	9.1
122	6	420	19	53	2.4
118	6	426	23	56	0.9
123	1	382	37	47	4.6
123	1	426	19	53	2.4

Note: Fluidized-bed gasifier, 610-mm diameter; flared steam at 121°C.

duction from biomass; however, additional losses would be incurred if a steam turbine generator were added for power production.

Curves illustrating typical fluidized-bed gasifier/boiler system performance are shown in Figures 2.12 through 2.14. These curves provide significant information for operating and controlling the gasification process. Important temperatures are shown in Figure 2.12. Operating bed temperature in the gasifier was maintained within ±25° of 760°C. If temperatures exceeded 845°C, agglomeration of bed particles was likely to occur, so fuel and air were discontinued if temperatures approached this value.

Temperatures in the freeboard area above the fluidized bed were used as an indication of gasifier performance. In start-up, vapor space temperatures were maintained lower than bed temperatures to prevent entering the combustion mode of operation. During longer-term operation, temperatures in the freeboard space will stabilize near bed temperatures.

Cyclone temperature in the experimental unit approximated the temperature of the gas entering the mixing-chamber. For adequate burning of the LCV gas, cyclone temperatures needed to be above 640°C.

Temperatures of the flame in the burner were not measured but temperatures of the gases exiting the flame tube of the boiler were measured and identified as midboiler temperatures. Those of the gases after they exited the small boiler tubes were identified as boiler-exit temperatures.

The differential between mid-boiler and boiler-exit temperatures

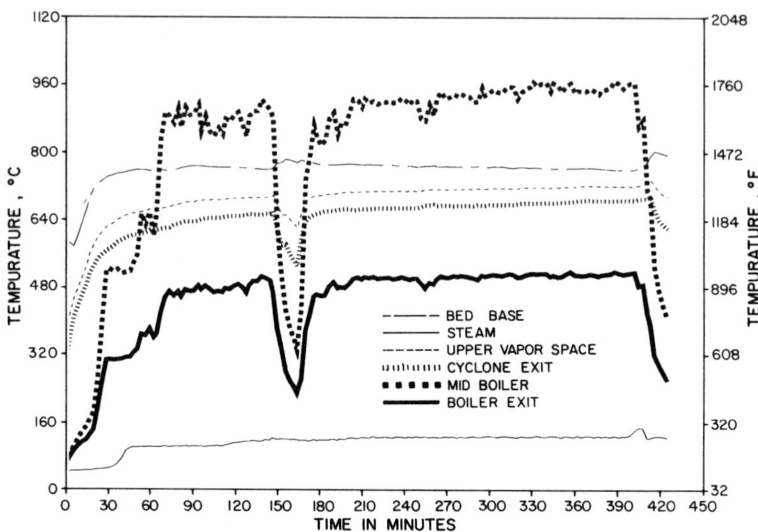

FIG. 2.12. Typical temperatures within fluidized-bed steam-producing system after initiating gin trash fuel feed.

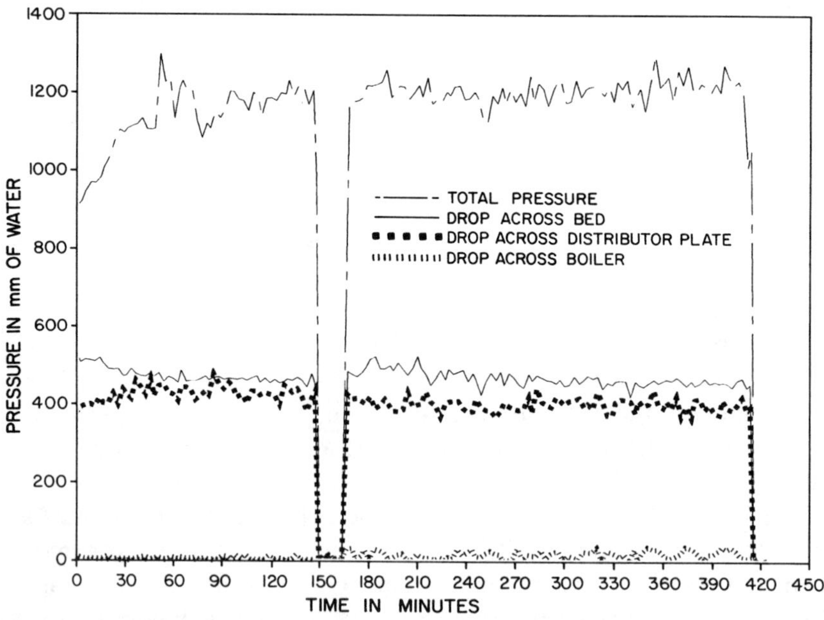

FIG. 2.13. Typical pressures within fluidized-bed steam-producing system after initiating gin trash fuel feed.

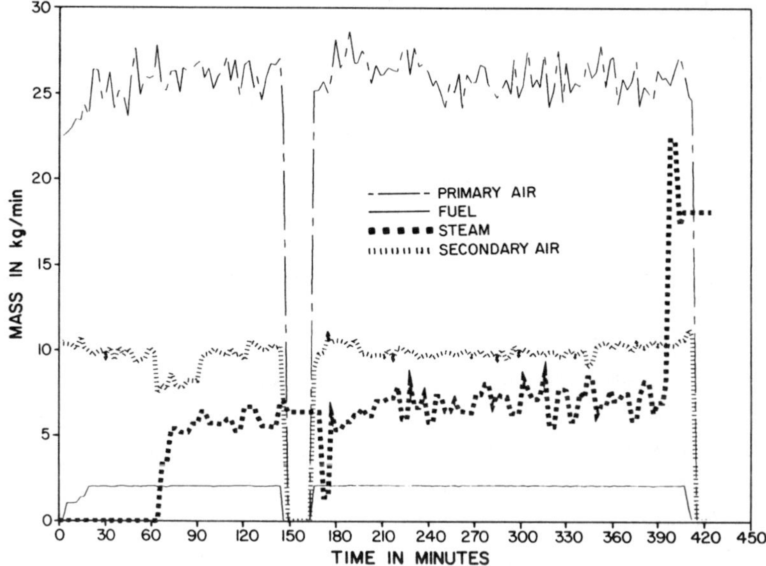

FIG. 2.14. Typical mass flow rates within fluidized-bed steam-producing system after initiating gin trash fuel feed.

gives an indication of the boiler heat transfer characteristics and efficiency, as seen in Figure 2.12. Boiler-exit temperatures were higher than normal for a boiler because of the long burning time of LCV gas and particle deposits in the boiler tubes. These were limitations of the experimental unit and can be improved in future designs; they are, however, important considerations in designing a system.

Pressures were monitored at various points within the system to determine operating parameters. These pressures are illustrated in Figure 2.13. Pressure drop across a laminar flow element was used for measuring primary air entering the gasifier.

Differential pressure across the bed was used as an indication of fluidization and the amount of bed material present. As airflow increases through the bed, pressure drop across the bed increases until fluidization occurs and then remains almost constant, as long as bed height remains constant. When major reductions in differential bed pressures occurred during operation, bed material was probably being entrained and removed from the system.

Differential pressure across the boiler gives an indication of any buildup on boiler tubes. Boiler deposits are partly a function of the quantity of ash not removed from the gas before it is burned; therefore, it is desirable to remove as much of the ash as possible. On this research facility the removal of particulate matter was inadequate, and deposition occurred in the boiler. However, two-stage cyclones were evaluated in a smaller gasifier, and a design model was developed to achieve satisfactory cleanup of the gas. This type of cleanup system is needed in order for the gas to be combusted effectively.

Mass flow rates of primary air, secondary air, fuel, and steam are shown in Figure 2.14. Actual rates of primary airflow through the gasifier are less than those shown in the curve, because significant leakage occurred through the fuel-feed rotary air lock. Leakage of air through the rotary air lock was a major source of error in determining airflow entering the gasifier, and failure to recognize this could lead to misinterpretation of air-to-fuel ratios as obtained from Figure 2.14. Air leaked through the feed system with every revolution of the rotary air lock because of different pressures inside and outside the system. This, in addition to wear in the unit, made it impossible to use measured airflow rates to determine air-to-fuel ratios.

Relative quantities of steam produced per unit of fuel can be observed by comparing curves for mass of steam to mass of fuel. From 2 to 3.7 kg of steam was produced from each kg of fuel in the typical test shown. This steam was flared to the atmosphere, so no pressure other than that of the exhaust system was maintained during the tests.

The boiler was rated at 545 kg of steam per hour for natural gas fuel. It was a fire-tube, dual-pass design; in interpreting the efficiencies measured, one should keep in mind that other designs may prove more efficient. In the selection of a boiler, gas velocities produced by burning the LCV gas should be a consideration, but the measurements shown here provide an estimate of the fuel needed to produce power.

The combustible components of the LCV gas sampled during this test are shown in Figures 2.15 and 2.16. A gas chromatograph was used to evaluate the samples, and heating values are shown in Figure 2.17. Gas heating values for this test ranged from 5,250 kJ/m^3 to 6,750 kJ/m^3. The composition is shown to vary some with time and is primarily influenced by bed temperature and fuel-to-air ratio.

The composition of exhaust gas from the boiler is shown in Figure 2.18. The presence of hydrocarbons in these samples reveals occasional incomplete combustion of the LCV gas. Although the design of

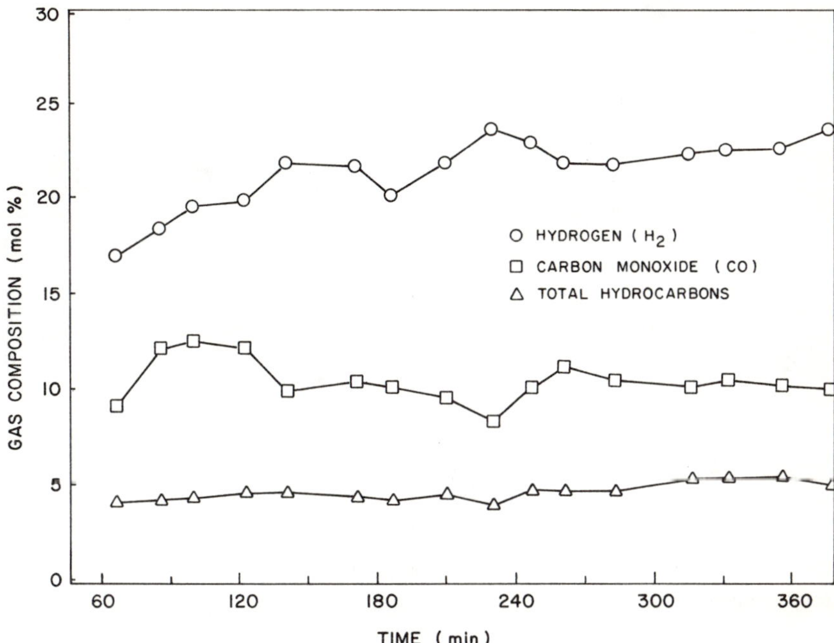

FIG. 2.15. Composition of product combustible gases as a function of time after initiating fuel feed. Gas samples were not taken during initial start-up period.

the mixing chamber and burner was not optimum, high overall conversion efficiencies were obtained. Refinement in the design of an LCV gas burner could further increase efficiencies.

Gas Cleanup. Fluidized-bed systems permit the use of a wide variety of biomass fuels but also produce high particulate levels in the gas. Most of the ash and some of the bed particles are entrained in the gas exiting a fluidized-bed gasifier, and studies have been made to evaluate cyclones and baghouses for removing particulates from LCV gas (LePori and Parnell, 1983).

A 305-mm-diameter fluidized-bed was used to evaluate a two-stage cyclone system and baghouse for removing particulates when using biomass fuels. A schematic of the test system is shown in Figure 2.19, and cyclone dimensions in Figure 2.20. The Teflon-coated fiber-

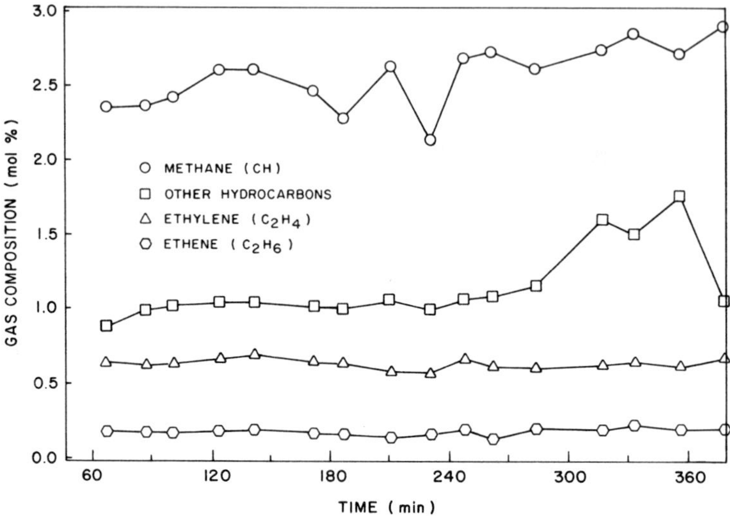

FIG. 2.16. Hydrocarbons composition in the gas product as a function of time after initiating fuel feed. Gas samples were not taken during initial start-up period.

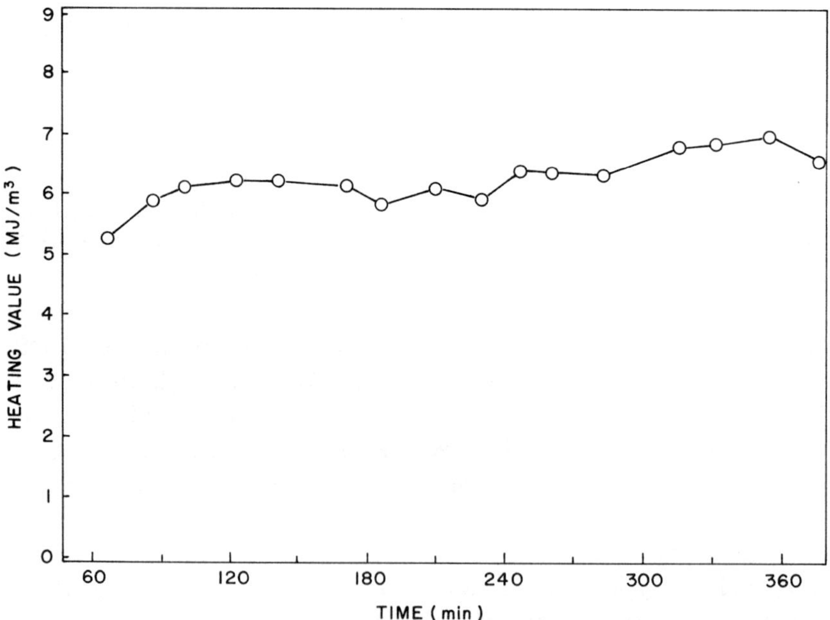

FIG. 2.17. Gas product heating value as a function of time after initiating fuel feed. Gas samples were not taken during initial start-up period.

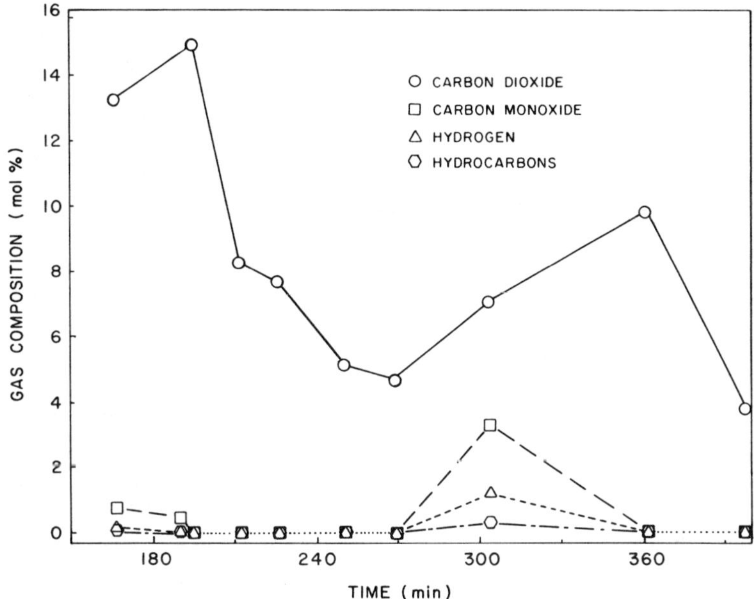

FIG. 2.18. Gas composition of exhaust from boiler as a function of time after initiating fuel feed. Gas samples were not taken during start-up period.

glass filter bags used in the baghouse have a maximum temperature rating of 260°C. Fuels having three ash levels were used in experimental evaluations: (1) manure, high ash; (2) cotton gin trash, medium ash; and (3) wood shavings, low ash.

Concentrations of particulates in the LCV gas exiting the gasifier were determined by summing the particulate concentrations measured at each sampling point. The effectiveness of the three cleaning devices was also determined from the inlet and outlet concentrations of each device.

The concentration of particulates remaining in the LCV gas after each stage of cleaning was of primary interest in these experiments. However, particle-size distributions and particle densities of the fractions removed by each cleaning device are also important for analyzing cleanup equipment performance and for designing more efficient devices. Particle sizes greater than 100 μm were removed from each sample, and the size distribution of the fraction less than 100 μm was determined using a Model TAII Coulter Counter. In addition, particle

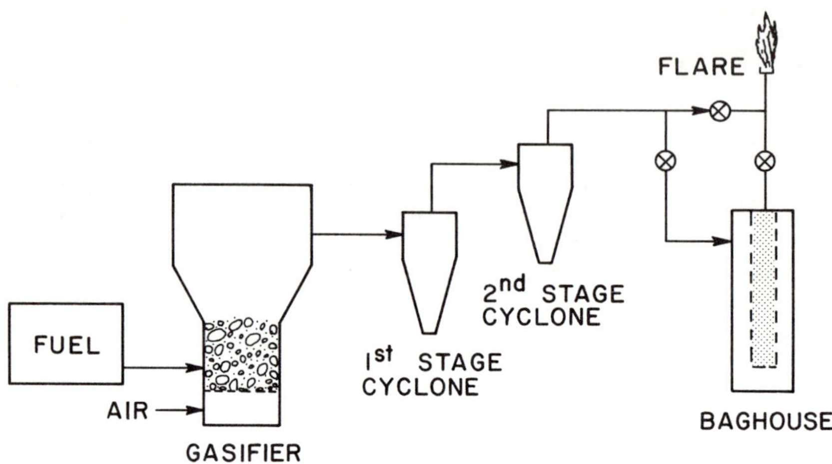

FIG. 2.19. Schematic showing equipment used in experimental determination of particulate concentration in LCV gas from a fluidized bed.

densities of representative ash samples were determined for the cleaning devices using an air pycnometer.

Results of experimental evalutions of particulates contained in the LCV gas from the 305-mm-diameter fluidized bed are summarized in Table 2.5. Gasifier operating conditions and gas composition for these experiments are shown in Table 2.6. The particulates consisted of unreacted carbon, ash, and elutriated bed material. The percentage of carbon in the collected material was evaluated, but the amount of bed material elutriated was not included.

The combined efficiencies of the cyclones are seen to be greater than 90 percent for all three fuels. Even though efficiencies are high, concentration leaving the secondary cyclone is still high for most uses. The level remaining after the secondary cyclone would be acceptable only when the gas is used as a fuel for furnaces or boilers. Even in these cases, a baghouse would probably be required after burning the gas as a final cleanup device to meet air quality standards.

Particulate Characteristics. The density of particles removed at each stage (see Table 2.7) is important for future design of abatement equipment and analysis of separation systems using computer simulation.

The ash captured by the primary cyclone contained a large per-

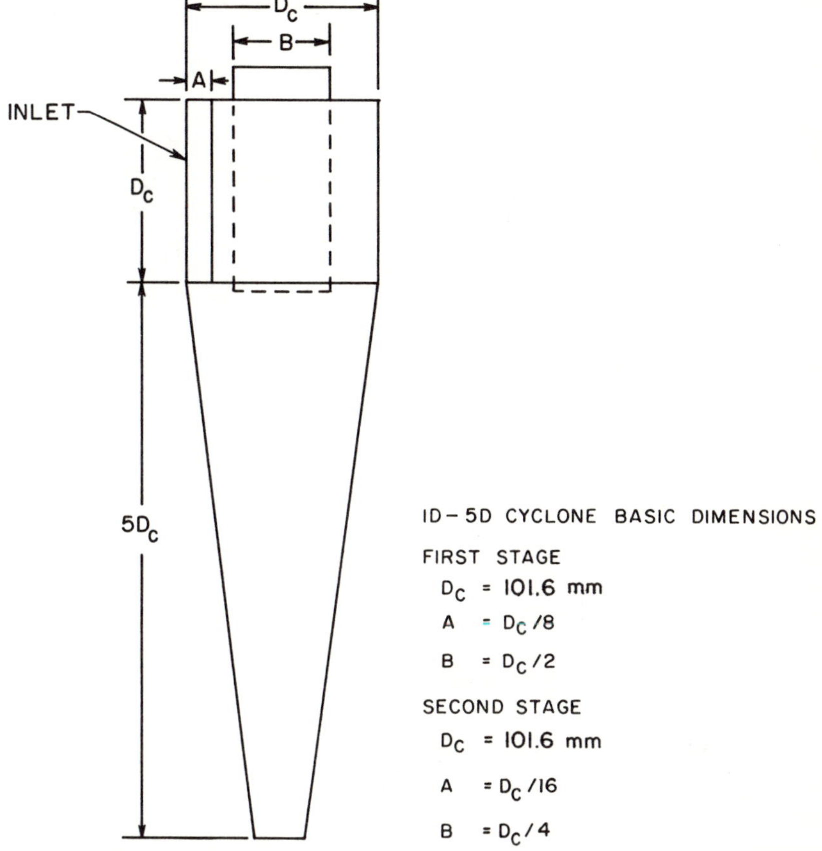

FIG. 2.20. Critical dimensions of cyclones used with a fluidized-bed gasifier for particulate evaluations.

centage of particulates larger than 100 μm. Sieving data indicated that the percentage of ash captured by the primary cyclone less than 100 μm for the three biomass materials was as follows: wood shavings (WS), 18.5 percent; cotton gin trash (CGT), 45.7 percent; cow manure (CM), 53.5 percent. These data suggest that the concept of using a primary cyclone to remove the larger particulates is important and should improve the effectiveness of the collection system. To illustrate this point, for ash contents of 5, 16, and 40 percent for WS, CGT, and CM, respectively, the primary cyclone would be required to remove at least

TABLE 2.5 Particulate Emissions Data Collected from LCV Gas

	Fuel				
	Gin Trash		Manure		Shavings
	Avg.	C.I. [a]	Avg.	C.I. [a]	Avg.
Primary cyclone					
Particulates collected					
Rate (g/m^3)	156	13.3	437	39.9	133
Size <100 μm (%)	45.7	4.5	53.5	8.8	18.5
Carbon (%)	19.5	3.7	5.9	3.2	13.5
Gas					
Flow (actual m^3/min)	1.8	0.08	1.48	0.22	1.64
Temperature (°C)	572	13	593	8	574
Secondary cyclone					
Particulates collected					
Rate (g/m^3)	10.4	3.07	16.2	8.50	1.46
Carbon (5)	42.0	—	11.4	—	62.5
Gas					
Flow (actual m^3/min)	1.53	0.07	1.29	0.21	1.38
Temperature (°C)	463	14	501	15	466

Leaving secondary cyclone[b]					
Particulates					
Rate (g/m^3)	4.59	2.25	3.28	0.97	8.80
Mg particulates/g fuel	8.1	—	4.6	—	15.3
Carbon (%)	56.4	—	—	—	—
Gas					
Flow (actual m^3/min)	1.08	0.05	0.87	0.18	0.92
LVC gas from fluidized bed					
Total particulates					
Rate (g/m^3)	171	33.6	456	41	14.3
Size <100 μm (%)	51.6	—	55.4	—	12.2
Carbon (%)	22.3	—	6.1	—	13.2
Percent of fuel feed	31.2	—	62.6	—	25.4
Cyclone efficiencies (%)					
Primary	92.6	—	95.7	—	91.1
Secondary	65.5	—	80.7	—	15.5
Combined	97.4	—	99.3	—	91.9

Note: Fluidized bed gasifier, 305-mm diameter.

(a) Confidence interval for 95% level of significance.

(b) Determined by either EPA Method 5 sampling or baghouse.

TABLE 2.6 Comparison of average gas composition and gasifier operating conditions

	Gin Trash	Manure	Shavings
Gasifier			
Bed temperature (°C)	762	762	756
Air flow (m^3/min)	0.40	0.42	0.39
Fuel feed rate (g/min)	306	339	300
Fuel-to-air ratio (kg/kg)	0.625	0.651	0.625
Carbon efficiency (%)	65.74	51.35	56.10
Cold gas efficiency (%)	45.61	28.74	46.29
Carbon saturation ratio	0.376	0.401	0.357
H/C ratio	1.217	1.011	1.056
Carbon in ash (%)	20.3	6.1	12.2
	Dry Mole Percent		
LCV gas			
Hydrogen	9.765	3.999	4.377
Methane	2.492	1.643	3.549
Ethylene	0.791	0.239	1.116
Ethane	0.200	0.031	0.246
Propane	0.180	0.292	0.279
Propylene	0.016	0.007	0.012
Carbon monoxide	11.203	5.744	12.098
Carbon dioxide	12.548	8.494	9.797
Nitrogen	56.629	71.786	60.348
Oxygen	6.536	7.771	8.188
Production rate (m^3/min)	0.56	0.47	0.51
Heating value (MJ/m^3)	4.25	2.22	4.41

Note: All three fuels used in a 305-mm fluidized-bed gasifier.

41, 83, and 186 grams of particulates larger than 100 μm for every 1,000 grams of biomass gasified.

Results of the Coulter Counter's determination of particle-size distribution (PSD) are shown in Table 2.8. It should be emphasized that the PSDs for ash captured by the primary cyclone are only for that fraction of captured ash that was less than 100 μm. These results are the average PSDs of all test samples analyzed and were converted to aerodynamic size values. The median size of ash particles captured by the

TABLE 2.7 Particle Densities for Captured Ash, Determined with an Air Pycnometer

	Ash Captured		
Fuel source	Primary Cyclone (g/cg)	Secondary Cyclone (g/cg)	Leaving Secondary Cyclone (g/cg)
Wood shavings	2.69	2.53	(a)
Gin trash	3.12	2.65	1.97
Cow manure	2.91	2.59	(a)

(a) Assumed to be the same as cotton gin trash. Concentrations and PSD analyses were obtained using EPA Method 5 sampling, and insufficient ash was available for measurement using an air pycnometer.

primary and secondary cyclones was reduced to 7.63 μm for wood shavings, 7.97 μm for cotton gin trash, and 8.48 μm for cow manure.

Cyclone Design Criteria. The cyclone cut diameter, which is the 50-percent collection point from the fractional efficiency curve, was approximated by determining the aerodynamic median diameter of captured particles less than 100 μm. The secondary cyclone had a cut diameter of approximately 8 μm and was approximately 60 percent efficient. The median diameter of the ash escaping the secondary cyclone was 8.34 μm, 4.90 μm, and 7.21 μm for WS, CGT, and CM, respectively. Only CGT suggested a much lower median diameter for particulates leaving the secondary cyclone. This could be a consequence of the measuring techniques: the ash escaping the secondary cyclone was captured by a bag filter at nearly 100 percent collection efficiency for CGT, while air sampling methodology was used for WS and CM. Higher confidence can be placed on the bag filter ash PSD data (Cheremisinoff and Cheremisinoff, 1973).

A lower aerodynamic median diameter for ash entrained in the gas leaving the secondary cyclone would suggest that the secondary cyclone is effectively removing particulates larger than the cut diameter, or that the ash size is being reduced by mechanical forces in the high-energy cyclone, or both.

Figures 2.21, 2.22, and 2.23 are plots of the differential particle-size distributions for the ash captured by the primary and secondary

TABLE 2.8 Statistical Analysis of Particle-Size Distributions Using the Coulter Counter

Geometric	Aerodynamic[a] Mean (μm)	Aerodynamic[a] Median (μm)	Aerodynamic[a] Mode (μm)	Std. Dev. (μm)	Skewness
Wood shavings					
Primary cyclone	24.9	25.9	47.6	2.08	0.87 negative
Secondary cyclone	7.95	7.63	7.31	1.69	1.09 positive
Leaving secondary cyclone	9.35	8.34	6.58	2.34	1.31 positive
Cotton gin trash					
Primary cyclone	18.7	19.2	20.5	2.05	0.93 negative
Secondary cyclone	8.27	7.97	7.59	1.78	1.09 positive
Leaving secondary cyclone	5.78	4.90	3.06	2.07	1.49 positive
Cow manure					
Primary cyclone	26.8	31.1	42.2	2.17	0.71 negative
Secondary cyclone	9.08	8.48	7.70	1.86	1.15 positive
Leaving secondary cyclone	8.03	7.21	5.88	2.15	1.21 positive

(a) Coulter Counter equivalent spherical size values were converted to aerodynamic by multiplying by $\sqrt{\text{particle density}}$.

cyclones and particulates leaving the secondary cyclone for each biomass feedstock. Note the similar PSDs for ash captured by the secondary cyclone for each biomass, suggesting that the secondary cyclone design factors the size characteristics of the ash captured. This is not the case for the primary cyclone, because of its extremely heavy particulate loading and the large variation in initial size characteristics of biomass ash from different biomass feedstocks.

Computer Model Verification. The particulate emissions measured from the two cyclones in these experiments were compared to predicted values from a cyclone design model developed at Texas A&M University (Parnell, Guzman, and Hickman, 1982). Comparisons of the measured and predicted values are shown in Table 2.9.

The values reveal that the primary cyclone is slightly more efficient and the secondary cyclone slightly less efficient than the model predicts. The measured and predicted emissions concentrations are in close agreement. On the basis of these tests, it appears that the model can be effectively used to optimize designs of cyclones for removal of particulates from LCV gas.

Tar Concentrations. Gas samples were taken from a 1.27-cm-diameter manifold provided on the exit line of the primary cyclone, using a vacuum pump. The samples first went through a set of four traps in an ice case, which were used to collect the condensable vapors.

The concentration of tars in the LCV gas produced from two biomass fuels is shown in Table 2.10. In addition, the average composition and heating values of the LCV gas are shown in Table 2.11. Tars in LCV gas are not a problem for some applications: for example, in a close-coupled boiler system, where the gas is not cooled, since it is used as a burner fuel. In such uses as fuel for an internal combustion engine, however, tars could be a severe problem and would need to be removed to prevent engine damage.

The amount of fuel input per unit-area of bed is also illustrated in Table 2.10. Larger fuel or energy input per unit-area decreases unit size for a given output. This could have a significant impact on the capital costs of a power-producing system.

Summary: Gasification

Though gasification is a more complex process than direct combustion, it offers an opportunity to convert biomass materials that

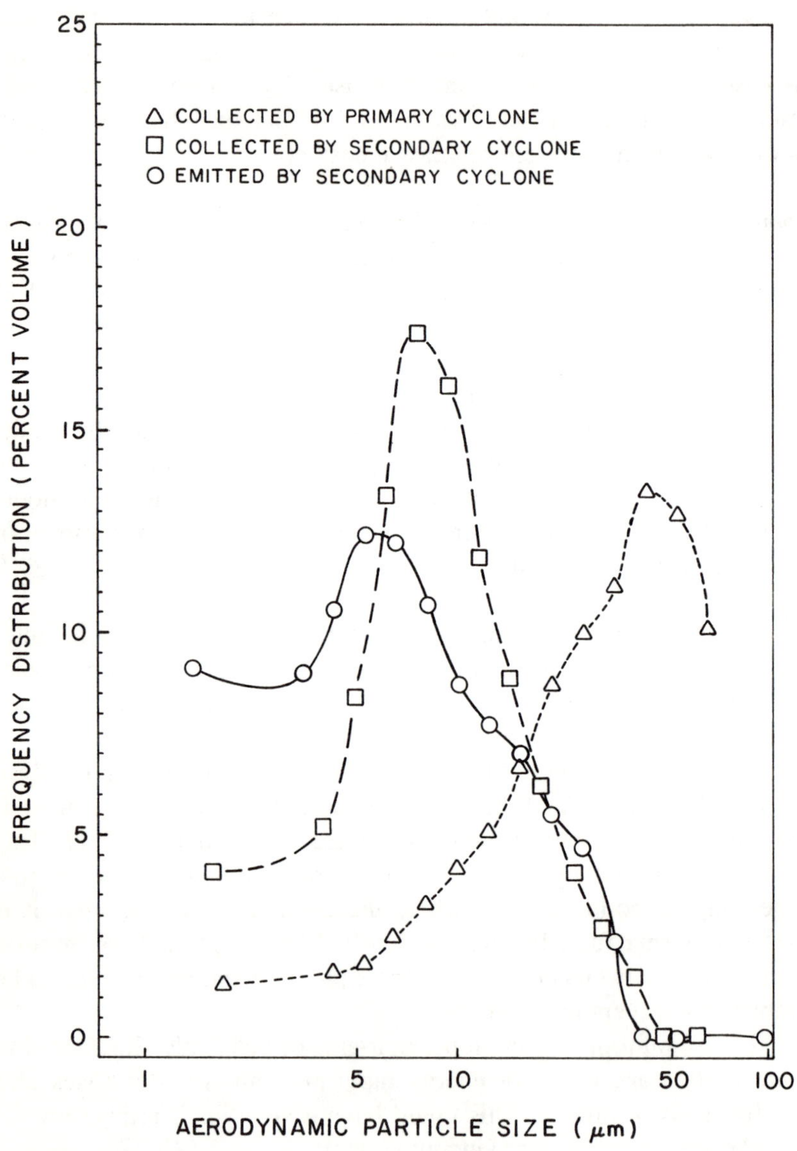

FIG. 2.21. Differential particle-size distributions for ash captured by primary (1st) and secondary (2nd) cyclones and particulate entrained in gas leaving secondary cyclone (3rd) for cow manure biomass fuel.

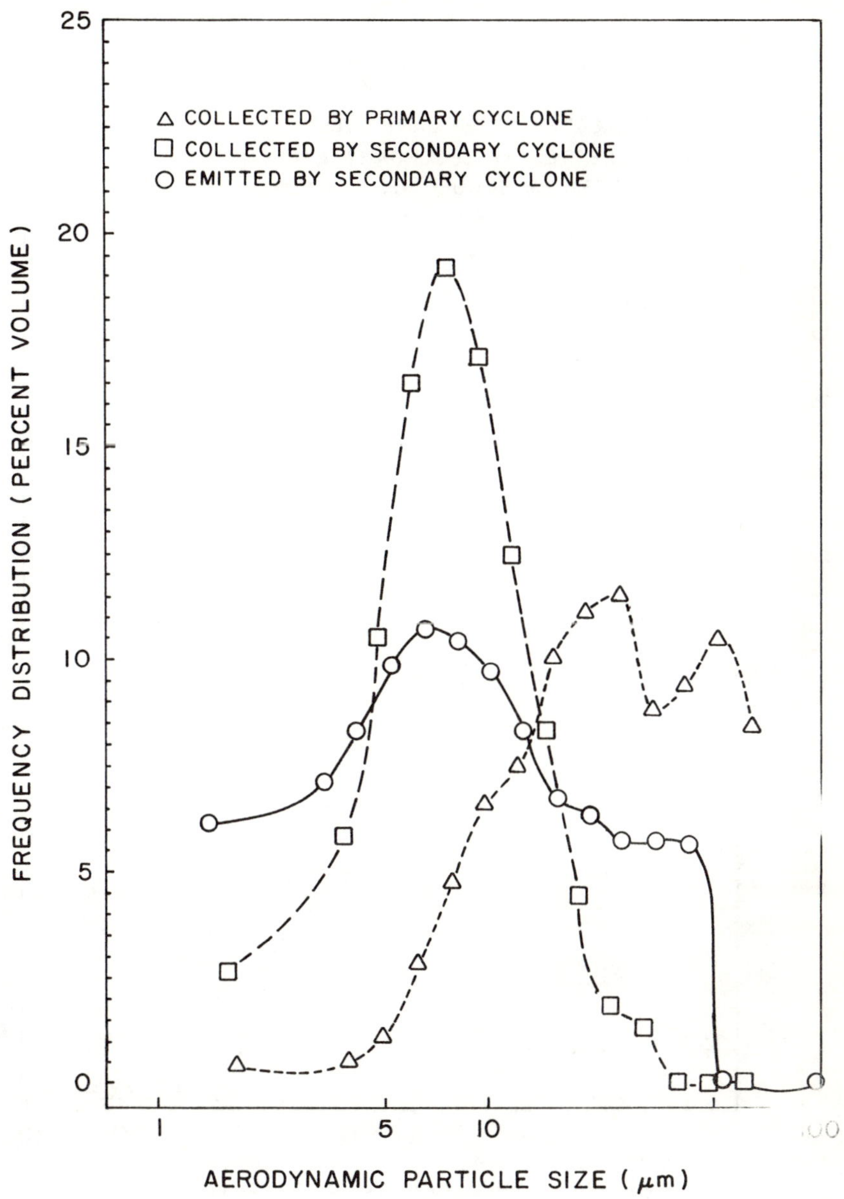

FIG. 2.22. Differential particle-size distributions for ash captured by primary (1st) and secondary (2nd) cyclones and particulate entrained in gas leaving secondary cyclone (3rd) for wood shavings biomass fuel.

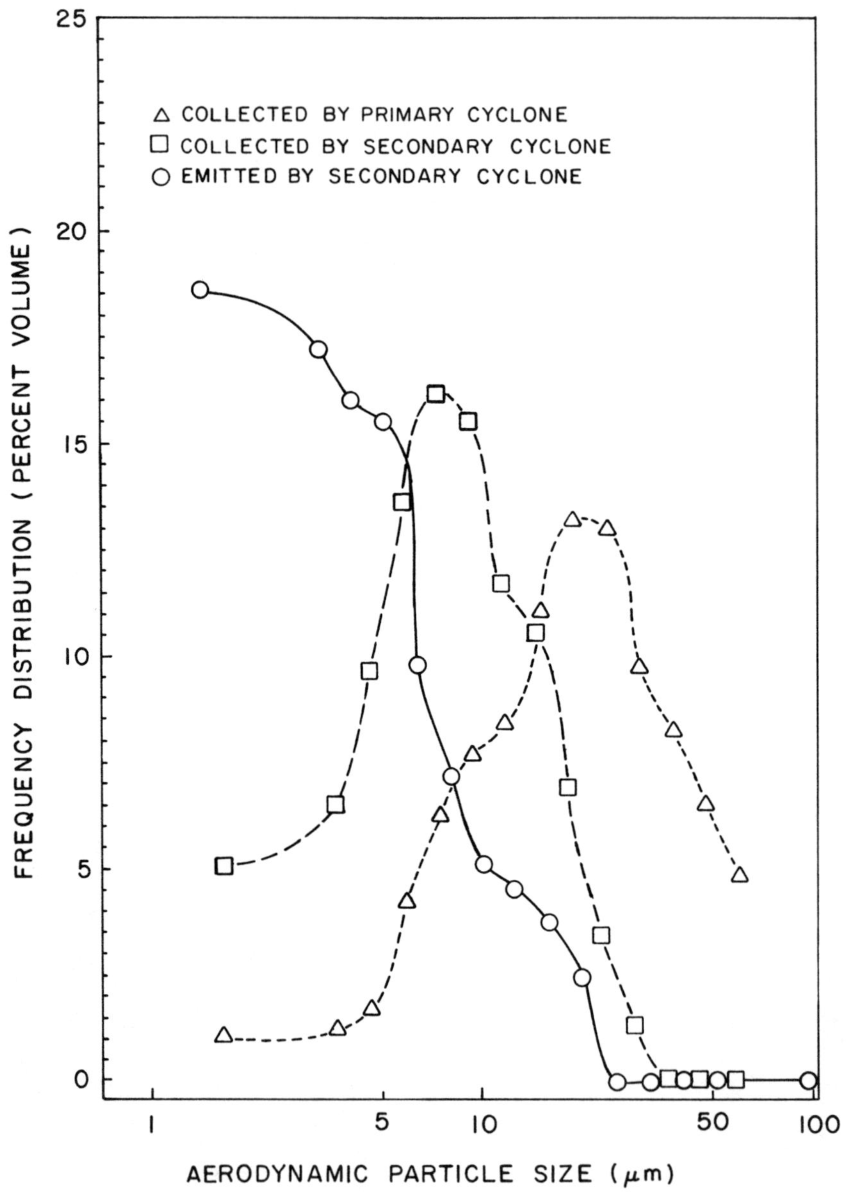

FIG. 2.23. Differential particle-size distributions for ash captured by primary (1st) and secondary (2nd) cyclones and particulate entrained in gas leaving secondary cyclone (3rd) for gin trash biomass fuel.

TABLE 2.9 Comparison of Average Measured and Predicted Values

	Primary Cyclone		Secondary Cyclone	
	Cotton Gin Trash	Cattle Feedlot Manure	Cotton Gin Trash	Cattle Feedlot Manure
Efficiency (%)				
^{x} measured[(a)]	84.1	92.4	65.6	80.7
^{s} measured[(b)]	4.6	2.3	22.8	8.8
^{x} predicted	79.4	85.0	81.9	85.3
^{s} predicted	1.0	1.2	7.0	1.6
Emitted concentration (g/m^3) actual				
^{y} measured	5.3	7.1	3.3	1.7
^{s} measured	2.2	2.9	2.4	0.5
^{y} predicted	6.6	12.1	0.9	1.0
^{s} predicted	2.1	2.2	0.2	0.3
Cut diameter (μm)				
^{z} measured	2.9			3.6
^{s} measured	0.7			1.4
^{z} predicted	3.7			1.5
^{s} predicted	0.03			0.5

(a) x, y, and z = variables of interest.
(b) s = sample standard deviation.

could not otherwise be used for fuel. The gasification process apparently controls the chemical reactions in such a way as to prevent fouling and slagging of system components. Separating the ash from LCV gas removes the chemicals that cause formation of eutectic compounds having low melting points. LCV gas can then be used in many secondary processes, free of the eutectics that can create operating problems in a high-temperature oxidizing environment. A proposed biomass steam-producing system based on the principles established in this research is illustrated in Figure 2.24. This provides a conversion system for many biomass fuels that could not be used by other thermochemical conversion techniques.

Pyrolysis

Pyrolysis is a thermochemical process through which lignocellulosic materials (biomass) can be converted to solid char, liquid tar, and

TABLE 2.10 Tar Concentration in LCV Gas (305-mm fluidized-bed gasifier)

Feedstock	Fuel Rate (kg/min)	Air Flow (m^3/min)	Fueling Rate (MJ/m^2/hr)	Bed Temp. (°C)	g of Tar/ m^3 gas	m^3 Gas/ min	mg Tar/ g Feed	g of Tar/ min
Cottonseed hulls	0.99	0.76	10,757.6	752	5.0	1.35	6.77	6.70
Cottonseed hulls	0.94	0.74	10,214.8	760	13.3 [a]	1.30 [a]	18.38 [a]	17.28 [a]
Cottonseed hulls	1.1	0.71	11,892.6	754	42.4	1.25	48.18	53.0
Cottonseed hulls	1.3	0.94	14,063.9	767	26.5	1.65	33.63	43.72
Cottonseed hulls	1.3	0.93	14,359.9	781	22.5	1.63	28.21	36.67
Cotton gin trash	0.45	0.59	4,809.6	785	1.9	1.04	4.47	2.01
Cotton gin trash	0.66	0.60	7,141.5	758	3.4	1.06	5.46	3.65
Cotton gin trash	0.93	0.83	9,910.7	776	3.0	1.47	4.76	4.41
Cotton gin trash	0.95	0.86	10,202.2	779	3.4	1.52	5.46	5.21

(a) A filter to catch the particles was provided before the ice traps, and extraction to those particles was not performed.

TABLE 2.11 Composition of LCV Gas (305-mm gasifier)

Components	Cottonseed hulls Avg. Mol %	Cottonseed hulls Std. Dev.	Cotton gin trash Avg. Mol %	Cotton gin trash Std. Dev.
Hydrogen	9.71	1.13	14.78	1.83
Carbon monoxide	15.51	0.85	19.43	2.59
Carbon dioxide	20.10	0.57	15.84	1.20
Methane	4.14	0.35	2.33	0.43
Ethylene	1.07	0.16	0.76	0.15
Ethene	0.39	0.04	0.23	0.04
Other hydrocarbons	2.70	0.11	1.20	0.18
Oxygen	1.17	0.33	0.50	0.22
Nitrogen	45.09	1.94	44.54	2.08
Gross heating value (MJ/m^3)	212.73		166.86	

gaseous fuel products. The process is characterized by the use of temperatures generally below 600°C and the absence of oxygen (or air-starve conditions). Although all three primary products can be recovered in pyrolysis, it is possible—depending on product needs—to enhance yields of any of the three products by selecting process conditions (Soltes and Elder, 1981; Soltes, Wiley, and Lin, 1981).

At temperatures above 200°C, lignocellulosic materials thermally degrade to produce gases, liquids (tars), and solids (chars) as primary products. Depending on reaction parameters such as heating rate, final temperature, and residence time at temperature, as well as particle size, moisture or ash content, presence or absence of air or oxygen, and so on, these primary products can undergo secondary reactions affecting the yield and quality of the final products. Under conditions present in gasification processes, with relatively long residence times at elevated temperatures and in the presence of oxygen, char and tar products are broken down and partially oxidized to yield a gaseous product—mainly carbon monoxide and hydrogen with some carbon dioxide and hydrocarbon content. On the other end of the thermochemical process continuum, char is produced in carbonization processes under conditions of long residence time (hours to weeks) at lower temperatures (generally not greater than 400°C) with restricted

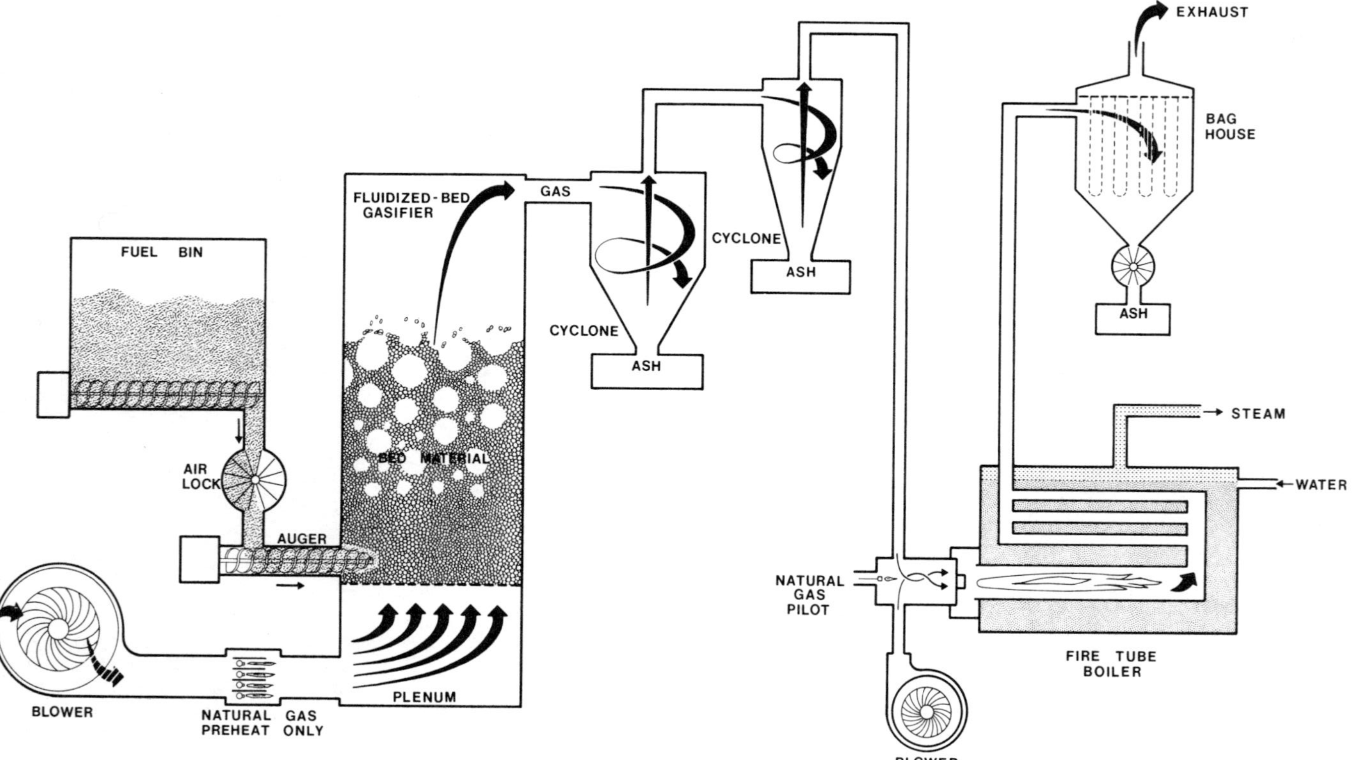

FIG. 2.24. Steam-producing system for fouling and slagging biomass fuels.

air supply. Given medium temperatures (450–600°C), rapid heating rates, abbreviated volatiles residence time (fractions of a second to minutes) at reaction temperatures, and oxygen-deficient or inert atmospheres, the primary tar product undergoes little secondary reaction; tar yield is thus enhanced.

Biomass pyrolysis research over the last decade has been process oriented. Very recently, interest has shifted toward "fast" pyrolysis (Milne, 1979; Soltes and Elder, 1978; Diebold, 1980), characterized by extremely rapid heat-up (for example, 10,000 degrees C per second) and very short residence times at temperature (for example, 30–40 milliseconds)—conditions that favor the production of olefins such as ethylene and propylene. There are suggestions (Milne, 1979; Diebold, 1980) that all pyrolysis short of the reaction conditions used in fast pyrolysis now be called "slow" pyrolysis, even though reactions can be fast and such suggestions further confuse the definition of biomass pyrolysis.

The problem in defining pyrolysis is that it is a process-oriented time used both for char and tar processes. Since the meanings of the terms *gasification* and *carbonization*—both product-oriented terms—are clear, it has been suggested (Soltes, Wiley, and Lin, 1981) that the term *tarification* be used for pyrolysis processes that favor tar (pyrolytic oil) production, to eliminate some confusion in the process-oriented term *pyrolysis*.

Opportunities for "slow" biomass pyrolysis relate directly to its inherent abilities in biomass tarification. The reaction parameters that enhance tar yields, the chemical composition of such tars, and some selected approaches to tar utilization are examined here.

Pyrolysis is at the same time humanity's oldest and newest thermochemical conversion process: old in its use for the production of pine tar, pitch, and charcoal; old in its use in wood-destructive distillation processes at the turn of the century for the production of organic chemicals; but new in some of its promise (Soltes and Elder, 1981).

Product Formation Mechanisms in Pyrolysis

Although there have been a number of reviews of cellulose and biomass pyrolysis kinetics and product-formation mechanisms (Milne, 1979; Browne, 1958; Roberts, 1970; Maa and Bailie, 1973; Williams, 1974; Broido, 1976; Shafizadeh and Groot, 1976; Lewellen, Peters, and Howard, 1976), biomass pyrolysis is still not completely understood.

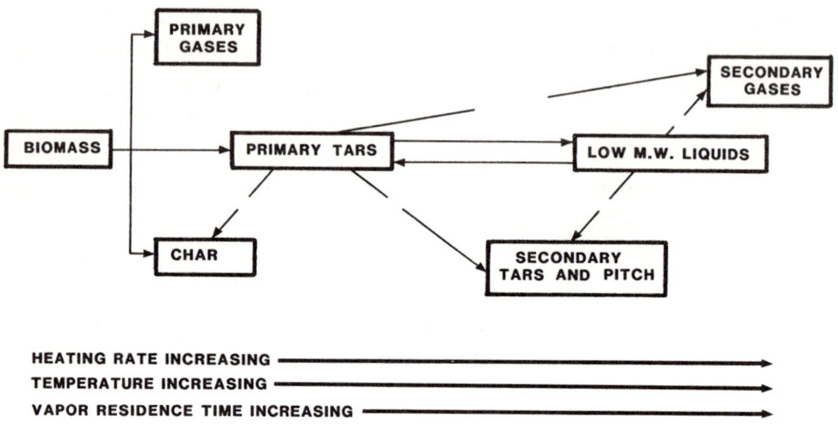

FIG. 2.25. Effects of heating rate, temperature, and vapor residence time at temperature on products formed in wood pyrolysis.

Part of the problem is that wood and other lignocellulosic materials, upon being subjected to pyrolysis conditions, behave much like a mixture of the three principal components: cellulose, lignin, and hemicellulose (Milne, 1979; Browne, 1958). Hemicelluloses decompose first, at 200–260°C, followed by cellulose at 240–350°C and lignin at 280–500°C (Roberts, 1970). Further, Milne (1979) concluded that (1) hemicellulose and holocellulose tend to yield less tar and more gas than cellulose; (2) lignin yields more aromatic substances than cellulose and hemicellulose; and (3) formic and acetic acids produced in hemicellulose pyrolysis may affect lignin and cellulose pyrolysis. Additionally, wood as a solid is subject to heat and mass transport effects in pyrolysis which are even more complex because wood is anisotropic. Thus, it is understandable that the kinetic analysis of wood pyrolysis is a difficult task. Dynamic TGA (thermogravimetric analysis) data, well developed for cellulose but less so for other components and wood itself, can nonetheless be used to predict pyrolysis data over a range of conditions; and they have engineering utility in the design and understanding of gasifiers and pyrolyzers having slow heating rates (Milne, 1979).

Considering the established sequence of pyrolysis for hemicelluloses, cellulose, and lignin components, one could argue that at any given elevated temperature, hemicelluloses and cellulose would be more thoroughly degraded than lignin, with lignin being the key component producing less degraded, higher-molecular-weight tar volatiles. However, most of the modeling work for pyrolysis mechanisms has been done on cellulose. Nevertheless, design considerations for enhancing tar production, gleaned from studies of cellulose pyrolysis, appear to be consistent with results obtained in process demonstration plants using wood and other biomass materials.

Process Design Guidelines

Excellent and detailed summaries of pyrolysis hardware are given in recent studies by Chatterjee (1981) and Robinson (1980). Although biomass pyrolysis is not strictly commercial, most of the pyrolysis systems have been demonstrated on a relatively large pilot scale and support the following design guidelines. (See also Figure 2.25).

Char. Low temperatures (200–400°C) and slow heating rates (hours to weeks) maximize the yield of char. Increasing residence time at the upper temperatures will result in char gasification reactions, which will reduce both char yield and volatiles content in the char. Counterflow conditions can decrease residence time at higher temperatures, thus favoring char production.

Fuel Gas. High temperatures (above 600°C) are needed to maximize carbon conversion to gaseous products. Extended gas residence time promotes cracking of condensable tars. Parallel flow may help to increase carbon conversion and decrease water and CO_2 content of the product, thus improving gas quality. In parallel-flow reactors, high moisture content may be desirable to increase carbon conversion via the water-gas reaction.

Tar. Tar yield is enhanced by employing medium temperatures (450–600°C), rapid heating rates, and abbreviated volatiles residence time. The combination of rapid heating rate and short residence time at upper temperatures does not permit secondary reactions of tar volatiles to gaseous and solid products. For parallel-flow reactors, a moisture content of about 20 percent appears to be optimum. In counter-flow reactors, some moisture may also be desirable to absorb heat from product volatiles.

Examples of counter-flow reactors include multiple-hearth fur-

TABLE 2.12 Typical Gross Biomass Pyrolytic Tar Composition

Components	%
Phenols, total	33.7
Neutrals	35.0
Acids	5.2
Water	9.7
Unextractables	15.1 [(a)]

(a) By difference.

naces and updraft gravitating beds. The gravitating bed is generally not ideal for high tar yields, since heating rate is slow. Rotary kilns may be parallel- or counter-flow operations. A characteristic of concurrent rotary kilns is relatively long residence time at high temperatures, resulting in low tar yields. Some parallel-flow reactors are downdraft gravitating beds and fluidized reactors. Fluidized-bed systems, with high heat transfer rates between sand and feed particles, exhibit relatively high heating rates conducive to high tar yields. Liquids, however, can be cracked by high pyrolysis temperatures and low fluidization velocities (high residence time). Fluidized-bed systems using lower process temperatures and higher fluidization rates should be good tar producers (Soltes and Elder, 1981).

Research on Thermochemical Tars

Tar Products. Depending on the process selected, liquid tar yield is variable but generally about 25 percent of dry feed, as reported for the Tech-Air pyrolysis process (Knight, Bowen, and Purdy, 1976). The tar, a complex mixture of organics, has potential as a feedstock for chemicals and fuels (Soltes and Elder, 1978; Elder, 1979; Elder and Soltes, 1979a, 1979b, 1980; Soltes, Wiley and Lin, 1981; Lin, 1981; Soltes, 1980a, 1980b, 1983a, 1983b). An analysis of the Tech-Air tar derived from the pyrolysis of pine bark and sawdust (Elder, 1979; Elder and Soltes, 1979a, 1979b) indicated that solvent extractable phenolics constitute approximately 13 percent of the tar in which phenol, creosol, guaiacol, dimethyl phenols, alkyl guaiacols, and eugenol are the major components (Table 2.12). Nonextractable, high-molecular-weight phenolics account for another 20 percent of the tar. Tar also contains volatile organic acids—which are responsible for its general cor-

TABLE 2.13 Comparison of Typical Fuel Oil and Biomass Tar

	No. 6 Fuel Oil	Biomass Pyrolytic Tar
Components (wt %)		
Carbon	85.7	59
Hydrogen	10.5	7
Oxygen	2.0	32
Sulfur	0.7–3.5	0.1
Nitrogen	—	1
Properties		
Density, g/mL	0.98	1.3
kJ/kg	42,333	24,656
kJ/L	41,473	33,167
Pour point, °C	18–29	32 [a]
Flash point, °C	66	111 [a]
Viscosity, SSU, 88°C	340	1150 [a]
Pumping temp., °C	46	71 [a]
Atomization temp., °C	105	116 [a]

(a) Pyrolytic tar containing 14% moisture.

rosivity (Lin, 1978)—as well as a variety of neutral components and nonvolatile materials (Soltes and Elder, 1981).

Fuels and Chemicals from Thermochemical Tars. Although reported attempts at tar utilization usually relate to direct combustion, tar is a poor liquid boiler fuel (Table 2.13) and a still poorer internal combustion engine fuel. It is viscous, gummy, not completely volatile, corrosive; it exhibits high oxygen content and does not mix with conventional fuels. Before tar can be used as an engine fuel, it must be processed to reduce viscosity and gumming tendencies, improve volatility, remove acidity, and lower oxygen content. As most of the undesirable properties are related to high-molecular-weight components and hydrophilicity, catalytic hydrocracking and hydrotreating were proposed.

Preliminary work in tar processing has been completed. Initially, catalysts were identified which exhibited specificities for converting a pine waste pyrolytic tar into a mixture of paraffinic fuel hydrocarbons.

TABLE 2.14 Catalyst Types Screened for Hydroprocessing Biomass Thermochemical Tars

5% Pd/alumina	0.5% Pd/alumina
5% Pd/carbon	0.5% Pt/alumina
	0.5% Re/alumina
5% Pt/alumina	
5% Pt/carbon	Harshaw Ni-4301
	Harshaw CoMo-0603
5% Re/alumina	Harshaw HT-400
5% Rh/alumina	
5% Ru/alumina	silica alumina
	$NiO-WO_3$ on silica alumina
Raney Ni	
$NiCO_3$	
UOP Lomax	ZrO_2 on alumina
UOP Unibon	

Source: Soltes (1983c).

Research was then extended to the processing of a wider range of biomass residues such as corncobs, sugar cane bagasse, cotton gin trash, rice hulls, and nut shells.

Hydroprocessing of Biomass Thermochemical Tars. Several types of catalysts were screened for this purpose (Table 2.14). Initial work, with 0.5 percent palladium and platinum catalysts on alumina with hydrogen-donor solvents such as tetralin and decalin, resulted in some hydrocracking and hydrotreating. Following these encouraging preliminary investigations, a series of experiments was conducted with higher concentrations of noble metals on various supports. Their use resulted in many improvements, and a number of these catalysts produced essentially the same results (Soltes, 1983c). There was more hydrogen uptake in the production of water (more oxygen removal), in the breakage of C-O bonds (removal of phenolic hydroxyl and methoxyl substituents, some cracking and reduction of acidity), in the removal of unsaturation (leading to products of light color), and in hydrocracking (higher yields of liquid product with lower viscosities). The hydrocarbons produced include alkylcyclohexanes, alkylbenzenes, and branched- and straight-chain hydrocarbons (C_5 to C_{30}) similar and identical to those found in conventional gasoline and diesel fuels.

Engine evaluations of fuels fractionated from the hydrotreated tar products, both diesel and gasoline types, have been very encouraging. Small spark-ignition engines will accept a light gasoline-like fraction directly. A mixture of 20 percent of this same fraction in indolene (a research gasoline) performed as well as 100 percent indolene in a research engine. A middle diesel-like fraction can be admixed up to 50 percent for use in diesel engines. Additional research into reforming these hydrocarbon mixtures should result in even better performance.

Tetralin and decalin work well as hydrogen-donor solvents but exhibit boiling points in the diesel fuel range. Thus, their use prevents both their recovery and the evaluation of the hydroprocessed product, or a distillative fraction thereof, as diesel fuel. More recently, the light ends of the hydrotreated product have been evaluated as solvent. The light ends contain the alkyl cyclohexanes, especially methyl cyclohexane, which appear to function just as well as decalin in hydrogen-donor activity in the system. These alkyl cyclohexanes are produced from the alkyl guaiacols present in the tar. Under the catalytic reaction conditions used, the phenolic hydroxyl and methoxyl moities of pine pyrolytic tar phenolics are split off the ring, followed by ring saturation to produce the alkyl cyclohexanes (Figure 2.26).

Compare Figure 2.27 for the volatiles of Tech-Air tar produced

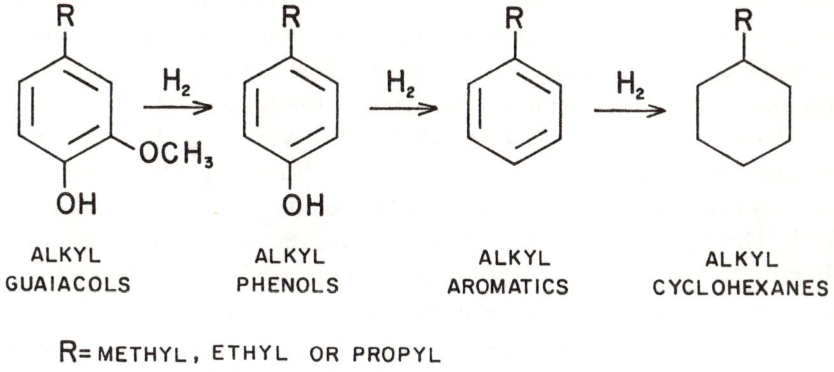

ALKYL GUAIACOLS $\xrightarrow{H_2}$ ALKYL PHENOLS $\xrightarrow{H_2}$ ALKYL AROMATICS $\xrightarrow{H_2}$ ALKYL CYCLOHEXANES

R= METHYL, ETHYL OR PROPYL

FIG. 2.26. Formation of alkyl cyclohexanes from alkyl guaiacols.

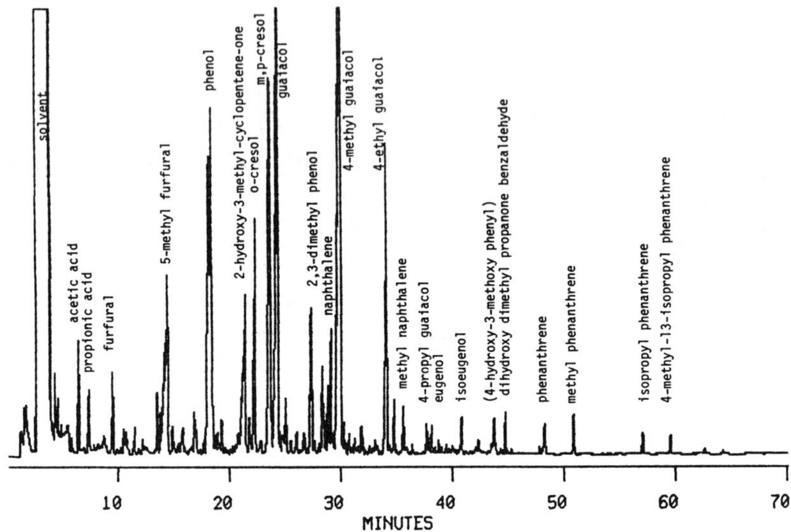

FIG. 2.27. Tech-Air pine waste tar.

via the pyrolysis of pine waste with Figure 2.28 for those of the hydroprocessed Tech-Air tar product. Note, in the former, the predominance of phenolics; in the latter, the appearance of hydrocarbons, alkyl aromatics, and alkyl cycohexanes.

Idealized Process Flow Sheet. An idealized process flow sheet for the conversion of biomass residues to gasoline and diesel hydrocarbon fuels is depicted in Figure 2.29. It is idealized in the sense that only a few subsystems have been studied (Soltes, 1983c, 1983d).

Biomass residues are first pyrolyzed to yield gas, tar, and char products (Figure 2.29a). The gas product, following Tech-Air practice, would probably be combusted to provide heat to dry incoming feed. The tar would be catalytically hydrotreated and hydrocracked (Figure 2.29b) to yield an oil product which would serve, when fractionated, to yield a middle-cut fuel feedstock. Light ends would be recycled for solvent purposes, and would aid as well in washing catalyst in catalyst recovery. Since there is typically no sulfur or other catalyst poison in biomass feedstocks, catalysts can enjoy long service life—

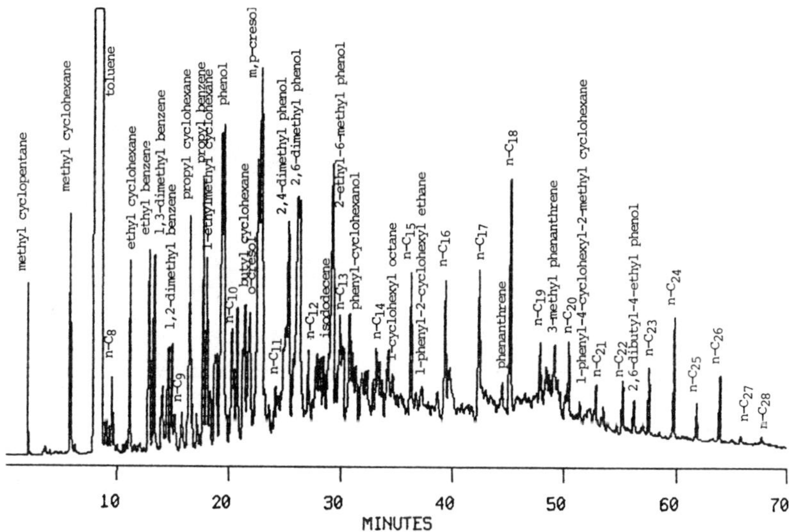

FIG. 2.28. Hydroprocessed Tech-Air pine waste tar, methyl cyclohexane solvent.

an important consideration in the use of noble metal catalysts. A small, heavier residual fraction can also be recycled for recracking as shown. Gasoline and diesel fuels would be produced by fractionation and blending of the middle-cut fuel feedstock (Figure 2.29c).

Further, the hydrogen content of the chars produced in biomass pyrolysis is sufficient to satisfy the hydrogen requirements of catalytic hydroprocessing; upon further study they may be available for such use through appropriate steam-char reactions. Activated carbons could be generated as by-products (Figure 2.29b). Energy requirements for both pyrolysis and subsequent processing may be derived from the gaseous products of pyrolysis. Thus, save for catalyst requirements, all raw material and energy needs to produce liquid hydrocarbon fuels, phenolics, and activated carbons can be supplied by a biomass feedstock.

To complete the process, an optional phenol recovery step is shown. Some catalysts and reaction conditions give mixed hydrocarbons and phenolic products. Besides the use of such phenolics for adhesives, several phenolics have been promoted as octane boosters (Elliott, 1981) and may additionally serve fuel interests.

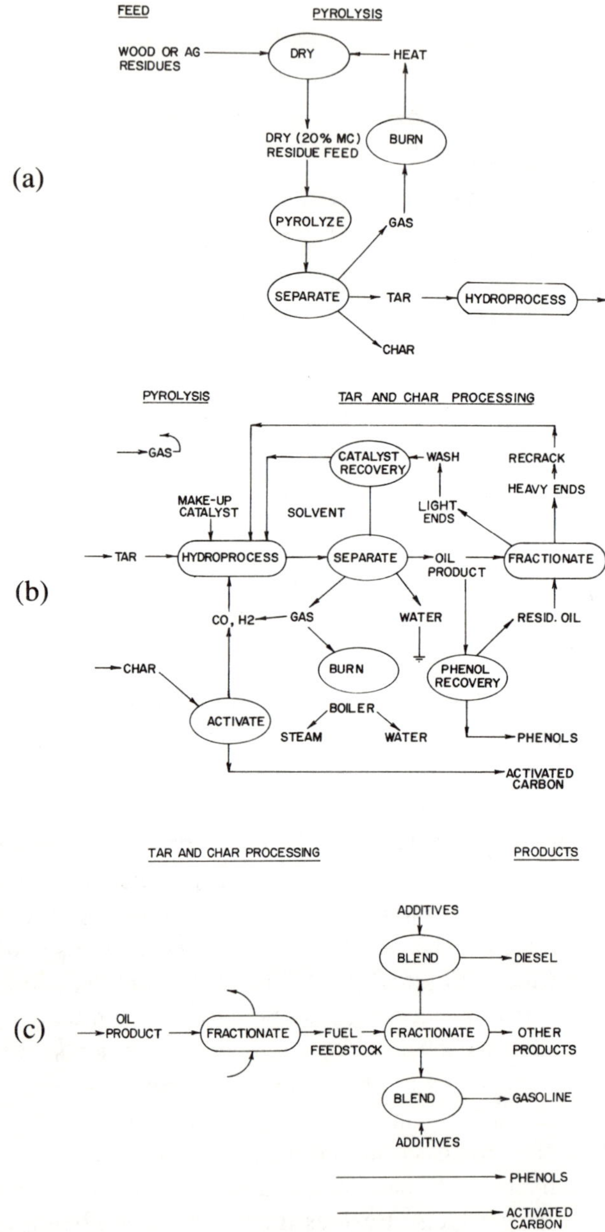

FIG. 2.29. Idealized process flow sheet for pyrolysis of wood and agricultural residues to yield gas, tar, and char primary products, which are subsequently processed to yield gasoline and diesel fuels, phenols, and activated carbon.

Process Flexibility. There is some compositional similarity in the volatile fractions of pine waste pyrolytic tar and corncob gasification tar (Soltes, Wiley, and Lin, 1981; Soltes, 1983c). This is not totally unexpected, since thermal degradation processes cannot distinguish between the cellulose, hemicellulose, and lignin composition of wood versus corncobs. Differences are primarily in the methoxyl substitution of phenolics present, as might have been predicted from the different lignin structures in the parent feedstocks. In subsequent processing, if the catalysts selected for hydrotreating are specific to carbon-oxygen bond cleavage, differences in methoxyl substitution should be eradicated.

It can now be reported that both the tars produced from a variety of agricultural residues and their hydrotreated products are indeed similar in composition. It appears that although the physical forms of such residues are different, tars produced via thermochemical processes exhibit many similarities in chemical composition, and that many differences in the composition of different tars are eradicated in catalytic hydroprocessing.

Figures 2.30 and 2.31 are representative gas chromatograms of the volatile constituents of tars derived from the updraft gasification of two types of agricultural residues—pecan shells and corncobs—and can be compared with the Tech-Air tar derived via wood waste pyrolysis (Figure 2.27). Similarities in phenolic composition are apparent.

Figures 2.32 and 2.33 are representative gas chromatograms of hydroprocessed tar products—peanut shells and sugar cane bagasse. Compare these chromatograms with each other and with that of the hydroprocessed Tech-Air tar in Figure 2.28. Note that the hydrotreated products contain alkyl aromatics, such as toluene and ethyl benzene, which are currently used by the petroleum industry as octane boosters for unleaded gasolines.

The similarities in both sets of chromatograms suggest that thermochemical conversion of biomass with subsequent hydrotreating of the tars may be a somewhat universal system for producing similar products from dissimilar biomass feedstocks. Although there has been much process research demonstration activity in biomass pyrolysis, and although pyrolysis has been promoted as a high-efficiency conversion process, there has been no implementation on a commercial scale. The major problem may be effective utilization of all three primary products; the process described offers one solution.

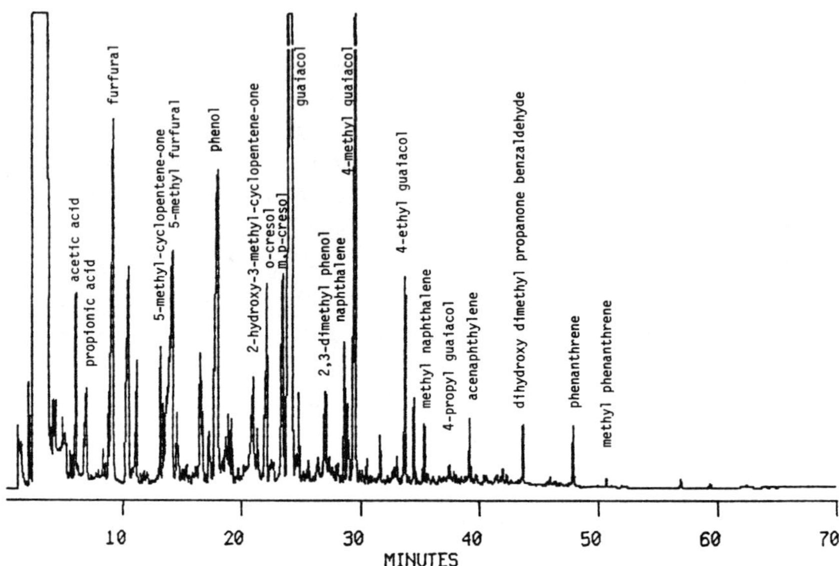

FIG. 2.30. Pecan shell tar.

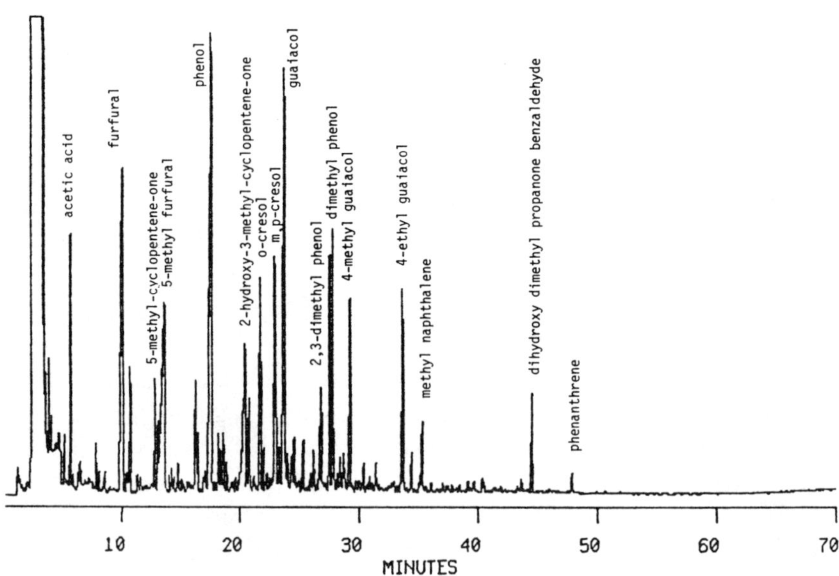

FIG. 2.31. Corncob tar.

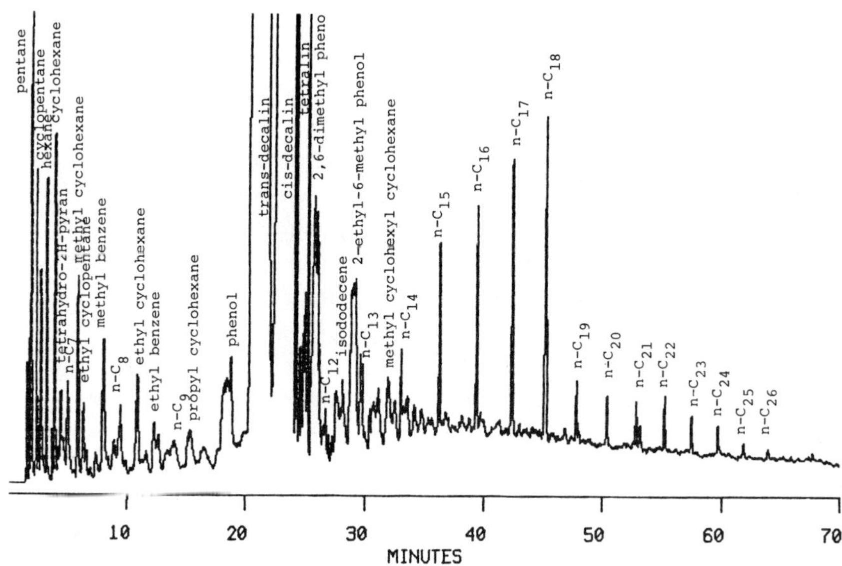

FIG. 2.32. Hydroprocessed peanut shell tar, decalin solvent.

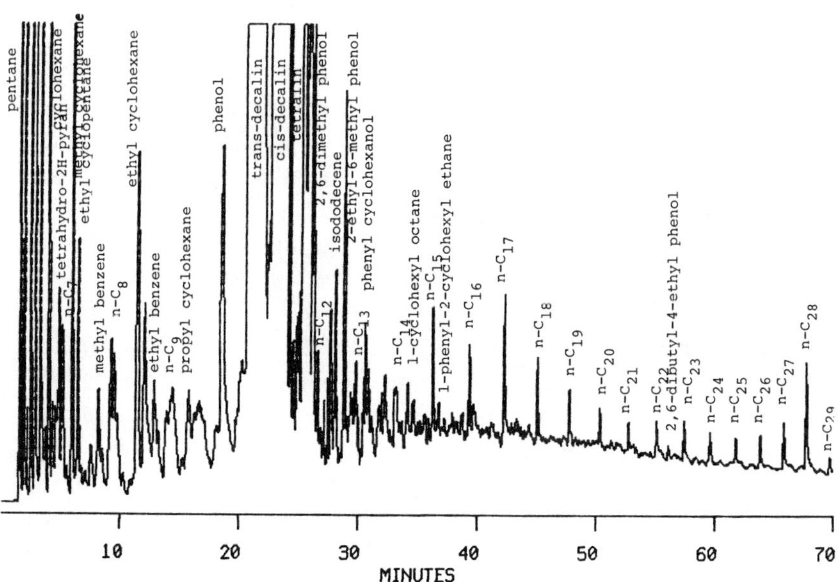

FIG. 2.33. Hydroprocessed sugar cane bagasse tar, decalin solvent.

Summary: Pyrolysis

As opposed to biomass combustion and gasification, for which tar production is undesirable, the major attribute of pyrolysis may be its use as a tarification reaction. Pyrolysis affords a one-step route to production of liquid tar from biomass. Examination of the chemical and physical properties of tars produced from a variety of biomass residue feedstocks shows that these tars are complex in chemical composition and exhibit less than desirable properties for use as liquid fuels. Nevertheless, tars can be processed to yield useful chemical products and energy. Based on research with the hydroprocessing of biomass tars, an idealized flow sheet has been developed which shows the processes for obtaining liquid hydrocarbon gasoline and diesel fuels, phenols, and activated carbon from the pyrolysis of wood and agricultural residues.

Uses for Biomass Thermochemical Processes

Solid Fuels from Biomass Residues

The most common form of biomass pyrolysis is carbonization: that is, a process to produce char. Char is produced in carbonization processes under conditions of long residence time (hours to weeks) at lower temperatures (generally not greater than 400°C) while restricting air supply. Char as a fuel is superior to wood or agricultural residues, and its production can be afforded in very simple systems (Soltes and Elder, 1981). In primitive carbonization systems, the gas and tar products are partially combusted to provide heat as the driving force for the conversion process, but to some extent are allowed to escape into the atmosphere. Even under such conditions, chars can be produced in 20 percent mass yield and—because of the higher calorific value of char compared with biomass—in some 30 percent energy yield.

Chars, then, are easily produced, are more energy intense than their parent biomass forms, and have the considerable advantage of being essentially smokeless; thus, they are heavily used as fuels for some cooking and heating purposes in developing countries not blessed with fossil fuels or the resources to purchase them. But the heavy usage of wood in charcoal-based economies in many parts of the world, coupled with unsound agricultural practices, is responsible for losses, sometimes irreparable, of forested areas (Soltes, 1980b). It is not difficult to increase char yields in more structured carbonization retorts, and it is tempting to suggest that efforts to assist and educate such societies in the use of more efficient processes should be expedited. However, a number of site-specific socioeconomic factors in rural areas of underdeveloped countries in charcoal-based economies make this transition difficult.

In other countries blessed with abundant biomass resources, notably Brazil, charcoal is produced in relatively sophisticated processes as a primary fuel for industrial energy needs, as well as in relatively primitive processes for cottage industries and home use, especially in remote rural areas. In more highly industrialized societies, using still more sophistication, carbonization can be an efficient process for the conversion of biomass into a more useful form of solid fuel. In the U.S., however, carbonization is relegated primarily to the production of charcoal briquets for home barbecues, even though wood charcoal has some interesting and useful properties that make it a unique raw material for the production of metallurgical and chemical products (Soltes and Elder, 1981).

Gaseous and Liquid Fuels in Agriculture

Agriculture (here including forestry) is a large consumer of energy; in any effort toward national energy independence, it would appear logical that agriculture should conserve energy and attempt to "grow" some or most of its energy requirements. To this end, various cultural and processing operations in forestry and agriculture have been changed to effect lower petroleum consumption. There has also been considerable activity in energy farming, in identifying fast-growth and energy-rich plant species and specifying agricultural acreage suited to their use. However, it may not be necessary to allocate large tracts of land to the production of energy crops. On-site biomass residues in and of many forestry and agricultural operations may be sufficient in energy content to satisfy the on-site energy requirements of such operations (Soltes and Wiley, 1978; Soltes, LePori, and Pollock, 1982; Lacewell et al., 1982). Nevertheless, biomass, whether specifically grown crops or available residue, is generally low in energy density, and its physical forms must be altered before fuel use can be considered.

On a local level in specific operations, and on a national level in the aggregate, shaft power produced via gas- and liquid-fueled engines is recognized as a major energy consumer in forestry and agriculture

(Soltes, LePori, and Pollock, 1982). Historically, we have generally favored gaseous fuels for stationary engines and refined liquid fuels for vehicular engines. There are several conversion options: gasification, pyrolysis, and hydrolysis/fermentation processes can change solid biomass materials into more useful gaseous and liquid fuel forms. Biomass gasification may serve stationary engine applications, and may be well suited to the scale of 2 to 200 tons of biomass feed per day generally required for local energy needs. For vehicular applications, however, liquid fuel options are more complex.

Several of these are given in Figure 2.34 (Soltes, 1983c). The first three—seed oils for diesel application, sugar crops and grain to produce fermentation alcohol for gasoline application—are relatively straightforward and have much appeal for small-scale, potentially on-farm implementation, but they suggest potential food-versus-fuel conflicts if implemented on scales of national energy significance (Soltes, 1980c; Soltes, Massey, and Murphey, 1981). Apart from questions raised about net energy contributions (Johnson, 1980; Weisz and Marshall, 1980), large substitutions of grain fermentation alcohol for gasoline could create a crisis-of-supply situation with drastic shifts in our use of agricultural land. The same problems in raw material demand might occur for the forest products industry if wood were used for either fermentation-derived ethanol or gasification-derived methanol on the scales once contemplated by the U.S. Department of Energy (1980). Simply, our national annual appetite for gasoline and diesel fuels is enormous compared with our agricultural and forestry production capacities.

The intermediary option, using residues to supply sugars for fermentation, circumvents potential resource conflicts but has to tackle the stubborn and, despite much research effort, yet unresolved problem of accessing sugars in lignocellulosics. Solving the accessibility problem of residues to improve digestibility by not only chemicals but enzymes and microorganisms must rank among the most important and challenging of endeavors in agricultural research (Soltes, 1980c, 1983e, 1983f).

Thermochemical Routes to Gaseous and Liquid Fuels

There are yet other options for liquid fuels. Thermochemical processing (gasification and pyrolysis) can accept a wide variety of biomass residue feedstocks, even dirty residues not amenable to chemical or biochemical processing, to yield gases and tars that can be used for

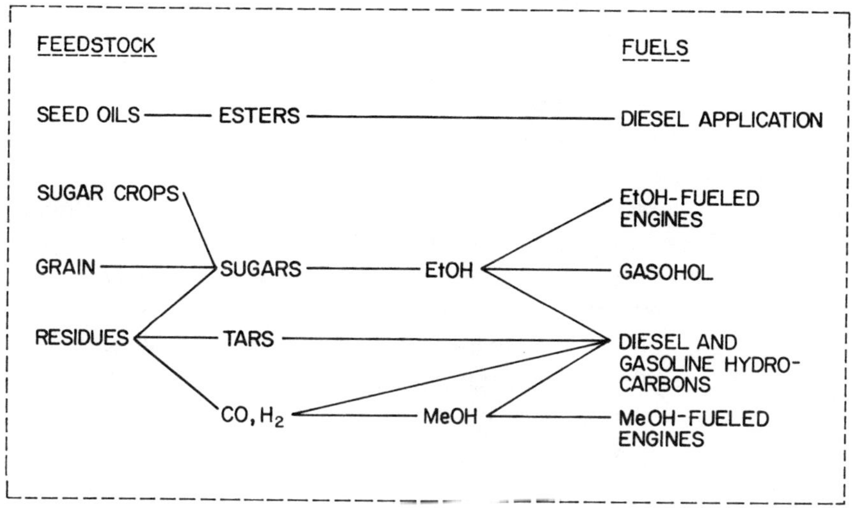

FIG. 2.34. Some options for engine fuels from biomass feedstocks.

liquid fuel synthesis. Gas produced via biomass residue gasification can be used in Fischer-Tropsch synthesis of paraffinic diesel hydrocarbons or, following reforming, gasoline (Kuester, 1981, 1983). Biomass residues can also be pyrolyzed to yield tars that can be catalytically hydrotreated into liquid hydrocarbon mixtures from which gasoline and diesel hydrocarbons can be fractionated. The advantages of thermochemical routes to engine fuels are that (1) residues instead of primary agricultural products can be used as feedstocks; (2) the processes are not restricted to certain types of feedstocks but are flexible in their use of essentially any biomass materials; (3) the processes are compatible with and similar to those currently used by the petroleum processing industry; (4) both gasoline and diesel fuels are produced; (5) the products are not substitute fuels, such as alcohol or seed oil, but the same hydrocarbons that are currently used in gasoline and diesel engines; and (6) with process variations, chemicals as well as fuels can be produced (Soltes, 1983g).

Summary

Thermochemical conversion in all forms allows the use of non-merchanted biomass residues for the generation of useful products. Thermochemical conversion in the form of combustion can assist us in meeting the energy needs of our forestry and agricultural operations. Other thermochemical conversion processes produce not only usable heat but a variety of useful, energy-dense fuels. These fuels, whether gases, tar, or char, can also be used as chemical feedstocks. Thermochemical conversion in the form of gasification can assist in adapting bulky biomass materials for use in efficient furnaces designed to burn gaseous fuel, and maybe in the generation of a versatile synthesis gas for the production of chemicals. Thermochemical conversion in the form of pyrolysis/carbonization assists those without access to fossil fuels by providing charcoal fuels, and in the form of pyrolysis/tarification may yet provide new approaches to oxychemicals and hydrocarbons for our material and vehicular engine fuel needs.

References

Attig, R. C., and A. F. Duzy (1969). Coal ash deposition studies and application to boiler design. *Proceedings, American Power Conference* 31:290–300.

Ballantyne, W. E., W. J. Huffman, L. M. Curran, and D. H. Stewart (1980). Energy recovery from municipal solid waste and sewage sludge using multisolid fluidized-bed combustion technology. In J. L. Jones and S. B. Radding, eds., *Thermal conversion of solid wastes and biomass*. ACS Symposium Series 130. Washington, D.C.: American Chemical Society.

Beck, Steven (1979). An overview of wood gasification. In K. E. Rogers, ed., *Proceedings, Tenth Texas Industrial Wood Seminar*, 20–37. Lufkin, Tex.: Texas Forest Products Laboratory, Texas Forest Service.

Broido, A. (1976). Kinetics of solid phase cellulose pyrolysis. In F. Shafizadeh et al., eds., *Thermal uses and properties of carbohydrates and lignins*. New York: Academic Press.

Browne, F. L. (1958). *Theories of combustion of wood and its control*. FPL Report No. 2136. Washington, D.C.: USDA Forest Service.

Chatterjee, A. K. (1981). *State-of-the-art review on pyrolysis of wood and agricultural biomass*. Final Report, Contract No. 53-319-R-0-206, AC Project P0380. Washington, D.C.: USDA Forest Service.

Cheremisinoff, P. N., and N. P. Cheremisinoff (1973). Fabric filters for dust collection. *Plant Engineering* 28 (June): 92–94.

Claar, P. W., II, W. F. Buchele, and S. J. Morley (1981). Development of a concentric-vortex agricultural-residue furnace. In B. F. Parker et al., eds., *Agricultural energy*, vol. 2: *Biomass energy and crop production*. St. Joseph, Mich.: American Society of Agricultural Engineers.

Craig, J. D. (1980). Performance and gas cleanup criterion for a cotton-gin waste fluidized-bed gasifier. M.S. thesis, Texas A&M University, College Station, Tex.

Culp, Archie W., Jr. (1979). *Principles of energy conversion*. New York: McGraw Hill.

Datin, Dennis L. (1983). Particulate cleanup of low-energy gas produced in a biomass fluidized-bed gasifier. Ph.D. dissertation, Texas A&M University, College Station, Tex.

Diebold, J. P., ed. (1980). *Proceedings, Specialists' Workshop on Fast Pyrolysis of Biomass*. SERI/CP-622-1096. Golden, Colo.: Solar Energy Research Institute.

Elder, T. J. (1979). The characterization and potential utilization of the phenolic compounds found in a pyrolytic oil. Ph.D. dissertation, Texas A&M University, College Station, Tex.

Elder, T. J., and E. J. Soltes (1979a). Further investigations into the composition and utility of a commercial wood pyrolysis oil. Paper presented at American Chemical Society National Meeting, Honolulu, Hawaii.

——— (1979b). Adhesive potentials of some phenolic constituents of pine pyrolytic oil. Paper presented at American Chemical Society Meeting, Washington, D.C.

——— (1980). Pyrolysis of lignocellulosic materials: Characterization of the phenolic constituents of a pine pyrolytic oil. *Wood & Fiber* 12:217.

Elliott, D. C. (1981). Description and utilization of product from direct liquefication of biomass. *Biotech. Bioeng. Symp.* 11:187.

Fung, D. P. C., and R. Graham (1980). The role of catalysis in wood gasification. In J. L. Jones and S. B. Radding, eds., *Thermal conversion of solid wastes and biomass*. ACS Symposium Series 130. Washington, D.C.: American Chemical Society.

Groves, J. D. (1979). Fluidized-bed gasification of agricultural residues. M.S. thesis, Texas A&M University, College Station, Tex.

Henry, J. E., H. M. Keener, and R. J. Anderson (1983). Performance of an automatically controlled fluidized-bed corncob combuster. Proceedings, Third Annual Solar and Biomass Workshop, 104–107. Tifton, Ga.: USDA, Southern Agriculture Energy Center.

Johnson, R. C. (1980). *Energy from biomass: The implications of gasohol from corn*. MP-81W6. McLean, Va.: MITRE Corp., Metrek Div.

Knight, J. A., M. D. Bowen, and K. R. Purdy (1976). Pyrolysis: A method for conversion of forestry wastes to useful fuels. Paper presented at Forest Products Research Society Meeting, Atlanta, Ga.

Kuester, J. L. (1981). Liquid hydrocarbon fuels from biomass. In D. L. Klass, ed., *Biomass as a nonfossil fuel source*. ACS Symposium Series 144. Washington, D.C.: American Chemical Society.

——— (1983). Catalytic conversion of biomass-derived synthesis gas to diesel fuel in a slurry reactor. *Proceedings, Third Annual Solar and Biomass Workshop*, 108–12. Tifton, Ga.: USDA Southern Agricultural Energy Center.

Lacewell, R. D., E. A. Hiler, S. M. Masud, and R. D. Kay (1982). *Assessment of integrated agricultural and energy production systems*. CEMR-MS4. College Station, Tex.: Center for Energy and Mineral Resources, Texas A&M University.

Lang, R. C. (1980). *Cotton gin trash incinerator air heat project*. Consultant Report P500-80-018. Sacramento, Calif.: California Energy Commission.

Lee, Y. N. (1981). Production of low-Btu gas from biomass. M.S. thesis, Texas A&M University, College Station, Tex.

LePori, W. A., R. G. Anthony, R. B. Griffin, T. C. Pollock, A. R. McFarland, and C. B. Parnell, Jr. (1981a). *Biomass energy conversion using fluidized-bed technology: Beyond the energy crisis*, 4:A297–A304. New York: Pergamon Press.

LePori, W. A., R. G. Anthony, T. R. Lalk, and J. D. Craig (1981b). Fluidized-bed combustion and gasification of biomass. In *Agricultural energy*, vol. 2: *Biomass energy and crop production*. ASAE Publication 4-81. St. Joseph, Mich.: American Society of Agricultural Engineers.

LePori, W. A., D. B. Carney, T. R. Lalk, and R. G. Anthony (1981c). *Process steam production from cotton-gin trash*. ASAE paper 81-3594. St. Joseph, Mich.: American Society of Agricultural Engineers.

LePori, W. A., and C. B. Parnell, Jr. (1983). *Clean-up of gases produced from gasification of agricultural biomass*. College Station, Tex.: Department of Agricultural Engineering, Texas A&M University.

Lewellen, P. C., W. A. Peters, and J. B. Howard (1976). Cellulose pyrolysis kinetics and char formation mechanism. In *16th International Symposium on Combustion*. Pittsburgh: Combustion Institute.

Lin, S.-C. K. (1978). Volatile constituents in a wood pyrolysis oil. M.S. thesis, Texas A&M University, College Station, Tex.

——— (1981). Hydrocarbons via catalytic hydrogen treatment of pin pyrolytic oil. Ph.D. dissertation, Texas A&M University, College Station, Tex.

Maa, P. S., and R. C. Bailey (1973). Influence of particle sizes and environmental conditions on high temperature pyrolysis of cellulosic material. *Combustion Sci. Technol.* 7:257.

Milne, T. (1979). Pyrolysis: The thermal behavior of biomass below 600°C. In *A survey of biomass gasification*, vol. 2: *Principles of gasification*. SERI/TR-33-239. Golden, Colo.: Solar Energy Research Institute.

Moreno, F. E. (1982). Fluidized-bed gasification of ginning wastes for energy: Two years with a full system. In *Proceedings, Cotton Gin Trash*

Alternatives. College Station, Tex.: Texas Agricultural Experiment Station, Texas A&M University.

Page, G. E. (1979). *Development of a burner for baled straw*. Department of Agricultural Engineering report. Corvallis, Ore.: Oregon State University.

Parnell, C. B. Jr., F. A. Guzman, and P. D. Hickman (1982). *Cyclone design methodology for agricultural processing*. ASAE paper 82-3582. St. Joseph, Mich.: American Society of Agricultural Engineers.

Payne, F. A., I. J. Ross, and J. N. Walker (1979). *Forced fed biomass gasification-combustion for drying grain*. ASAE paper No. 79-4546. St. Joseph, Mich.: American Society of Agricultural Engineers.

Payne, F. A., I. J. Ross, J. N. Walker, and R. S. Brashear (1981). Gasification-combustion of corncobs for drying corn. In *Agricultural energy*, vol. 2: *Biomass energy and crop production*. ASAE publication 4-81. St. Joseph, Mich.: American Society of Agricultural Engineers.

Probstein, R. F., and R. E. Hicks (1982). *Synthetic fuels*. New York: McGraw-Hill.

Raman, K. P., W. P. Walawender, Y. Shimizo, and L. T. Fan (1981). *Gasification of corn stover in a fluidized bed*. ASAE publication 4-81. St. Joseph, Mich.: American Society of Agricultural Engineers.

Reed, T. B. (1981). *Biomass gasification principles and technology*. Energy Review No. 67. Park Ridge, N.J.: Noyes Data Corp.

Richey, C. B., J. R. Barrett, and L. J. Klutz (1983). Development of a downdraft-channel gasifier furnace with mechanized fueling for crop drying. Paper presented at third annual Solar and Biomass Workshop, Atlanta, Ga.

Roberts, A. F. (1970). A review of kinetics data for the pyrolysis of wood and related substances. *Combustion Flame* 14:261.

Robinson, J. S., ed. (1980). *Fuels from biomass: Technology and feasibility*. Park Ridge, N.J.: Noyes Data Corp.

Sarkanen, K. V., D. A. Tillman, and E. C. Jahn (1982). *Progress in biomass conversion*, vol. 3. New York: Academic Press.

Schwieger, B. L. (1980). Power from wood. *Power* 124 (2): 51–532.

Shafizadeh, F., and W. F. Groot (1976). Combustion characteristics of cellulose fuels. In F. Shafizadeh et al., eds., *Thermal uses and properties of carbohydrates and lignins*. New York: Academic Press.

Smith, N., J. G. Riley, and D. F. Schaufler (1983). Processing and utilization of wood chip fuel. Proceedings, Third Annual Solar and Biomass Workshop, 270–73. Tifton, Ga.: USDA, Southern Agriculture Energy Center.

Soltes, E. J. (1980a). Pyrolysis of wood residues: A route to chemical and energy products for the forest products industry? *Tappi* 63:75.

——— (1980b). Thermal conversion of lignocellulosics: Retrospect and prospect. Paper presented at Intersciencia/JSST Seminar and Workshop on Materials for the Future, Kingston, Jamaica.

——— (1980c). *Alternate feedstocks for ethanol production: Small-scale production of ethanol from grain*. College Station, Tex.: Texas A&M University.

——— (1983a). Biomass thermal degradation tars as sources of chemicals and fuel hydrocarbons. In E. J. Soltes, ed., *Wood and agricultural residues: Research on use for feed, fuels and chemicals*, 477–87. New York: Academic Press.

——— (1983b). Thermochemical routes to chemicals, fuels and energy from forestry and agricultural residues. In W. A. Cote, Jr., ed., *Biomass utilization*, 537–52. New York: Plenum Press.

——— (1983c). Hydrocarbons from lignocellulosic materials. *Appl. Polymer. Symp*. 37:775.

——— (1983d). Fuels and chemicals from tars. In *Proceedings, Third Annual Solar and Biomass Workshop*, 116–21. Atlanta: USDA S&E Southern Agricultural Energy Center.

——— (1983e). *Wood and agricultural residues: Research on use for feed, fuels and chemicals*. New York: Academic Press.

——— (1983f). Cellulose: Elusive component of the plant cell wall. In W. A. Cote, Jr., ed., *Biomass utilization*, 271–98. New York: Plenum Press.

——— (1983g). Vehicular fuels and oxychemicals from biomass thermochemical tars. *Biotech. Bioeng. Symp*. 13:53.

Soltes, E. J., and T. J. Elder (1978). Thermal degradation routes to chemicals from wood. Paper presented at eighth World Forestry Congress, Jakarta, Indonesia.

——— 1981. Pyrolysis. In I. S. Goldstein, ed., *Organic chemicals from biomass*, 63. Boca Raton, Fla.: CRC Press.

Soltes, E. J., and A. T. Wiley (1978). Residue biomass resources of Texas: Boiler fuels for the Texas forest products industry. Unpublished MS.

Soltes, E. J., A. T. Wiley, and S.-C. K. Lin (1981). Biomass pyrolysis: Towards an understanding of its versatility and potentials. *Biotech. Bioeng. Symp*. 11:125.

Soltes, E. J., J. G. Massey, and W. K. Murphey (1981). Implications of selected wood use scenarios for the production of energy and industrial materials. *Biotech. Bioeng. Symp*. 11:3.

Soltes, E. J., W. A. Lepori, and T. C. Pollock (1982). Fluidized-bed energy technology for biomass conversion. *Biotech. Bioeng. Symp*. 12:15.

Sweeten, J. M., W. A. LePori, K. Annamalai, and C. B. Parnell (1983). Thermal conversion of cattle feedlot manure for energy production. Final report to Texas Cattle Feeders Association. Project No. 15090-6508.

College Station, Tex.: Texas Agricultural Experiment Station, Texas A&M University.

U.S. Department of Energy. (1980). The multi-year program plan for the office of technology development and utilization of the alcohol fuels group program. External comment draft, dated June 5.

Walawender, W. P. (1983). Fluid-bed gasification of crop residues, 112–15. Paper presented at third annual Solar and Biomass Workshop, Atlanta, Ga.

Wall, C. J., J. T. Graves, and E. J. Roberts (1975). How to burn salty sludges. *Chemical Engineering* 14 (April): 77–82.

Weisz, P. B., and J. F. Marshall (1980). *Fuels from biomass: A critical analysis of technology and economics*. New York: Marcel Dekker.

Williams, F. (1974). Chemical kinetics of pyrolysis. In P. L. Blackshear, ed., *Heat transfer in fires*. Washington, D.C.: Scripta Books.

Williams, R. O., and J. R. Goss (1979). An assessment of the gasification characteristics of some agricultural and forest industry residues using laboratory gasifier. In *Resource recovery and conservation*, 3:317–29. Amsterdam: Elsevier.

CHAPTER 3

Biological Conversion and Fuel Utilization: Anaerobic Digestion for Methane Production

JOHN M. SWEETEN and DONALD L. REDDELL

The concept of methane production appeals to many agricultural producers as a means of producing energy at the farm or feedlot. Crop residues, livestock and poultry manure, and food-processing wastes may be suitable substances (feedstocks) for methane production under some circumstances. Anaerobic digestion of these wastes, one form of bioconversion, offers four advantages: (1) the organic waste is converted into methane gas, which can be used as a direct energy source; (2) a high-protein livestock feed is produced, which can be used as an indirect energy source; (3) the mass of organic waste is reduced, thereby decreasing a potential pollution problem; and (4) the digester effluent can be utilized as a fertilizer.

A methane production system consists of the following steps (listed with some of the procedures and equipment required):

(1) feedstock handling—collection, hauling, and storage;
(2) feedstock preparation—grinding, dilution, mixing, preheating, and pumping;
(3) anaerobic digestion—internal agitation and heat exchange equipment;
(4) gas collection—conveyance pipe, pressure regulator, and condensate traps;
(5) gas scrubbing—removal of carbon dioxide, hydrogen sulfide, and moisture;
(6) gas utilization—compression, storage, marketing, or on-site utilization for heat or electricity;
(7) digested slurry handling—animal feed recovery (screening, centrifugation, and drying), or fertilizer recovery (storage pond and irrigation system).

The major components of an anaerobic digestion system are illustrated in Figure 3.1.

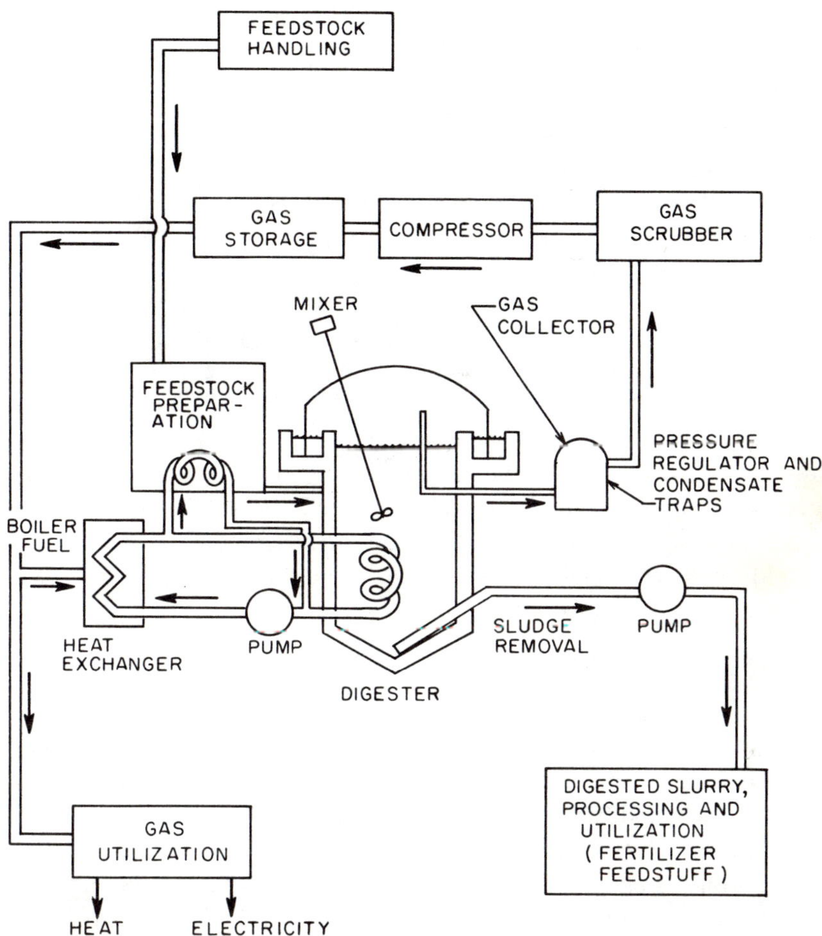

FIG. 3.1. Anaerobic digester components.

The Basic Process

Methane (CH_4), the primary component of natural gas, has a heating value of 37.3 MJ/m^3. Bacterial degradation of organic matter under anaerobic conditions (without oxygen) releases a mixture of gases (biogas), which usually consists of 50 to 60 percent methane, 40 to 50 percent carbon dioxide, and numerous trace gases—such as hydrogen sulfide—that together constitute 0 to 1 percent of the gas volume. This gives raw biogas an energy content of only 19 to 22 MJ/m^3. Carbon dioxide and trace gases can be removed by chemical means, yielding gas of pipeline quality (37 MJ/m^3).

Anaerobic digestion of organic residues and wastes is traditionally viewed as a two-phase process. In the first phase, bacteria degrade organic solids into organic acids. In the second phase, which occurs simultaneously, bacteria convert these organic acids into methane, carbon dioxide, and water.

More recent research has demonstrated that a three-stage process may be involved (Hashimoto et al., 1979a). The first stage consists of bacterial hydrolysis of carbohydrates, protein, and lipids, followed by fermentation of these products to fatty acids, hydrogen, and carbon dioxide. The second bacterial group is thought to produce acetates, hydrogen, and carbon dioxide from the fatty acids generated in the first stage. The third state involves methanogenic bacteria, which utilize the acetate, hydrogen, and carbon dioxide produced in the first two stages to yield methane (CH_4) and carbon dioxide (CO_2).

Benefits and Limitations

Anaerobic digestion has numerous attributes as an alternative energy production process. Foremost is the production of medium- to high-quality gaseous fuel containing, or consisting of, methane. A coproduct, carbon dioxide, may offer economic potential in some instances. A second major advantage is the retention of the feedstock's nutrient content and even, in the case of nitrogen, the possible upgrading of nutrient quality. Indeed, while the nutrient content remains constant, its concentration as a percentage of dry matter is increased 25 to 60 percent through the reduction of organic solids. Recovery of high-protein animal feeds is a third major advantage. Where wastes are used as feedstock, a fourth major advantage is bacterial stabilization of organic matter through reduction of volatile solids, by 35 to 65 percent in most cases, resulting in lowered odor potential and reduced size (and hence cost) of successive treatment system components. Secondary benefits of anaerobic digestion may include encouraged improvement in waste handling practices, through more frequent or thorough collection and lowered water consumption.

The limitations of the process include the capital cost of anaerobic digestion systems and gas-scrubbing equipment, and frequent difficulty in matching the methane gas output to erratic patterns of energy demand. Cost trade-offs exist between the expense of gas storage capacity versus conventional fuel demand and consumption charges. Where wastes are used as feedstocks, concerns for quality may preclude certain efficient collection practices such as flush systems, which create low solids contents, and infrequent scraping of manure from open, dirt-surfaced feedlots, which can yield high ash contents and biodegraded manure. The liquefaction of solid agricultural residues or wastes for methane production may complicate and delimit their ultimate disposal on land.

Terminology

In this discussion, the following terminology is used to describe the waste characteristics and the theory of anaerobic digestion.

Biochemical oxygen demand (BOD) is an indirect measure of the concentration of biologically degradable material present in an organic waste. BOD is the amount of oxygen required by bacteria while stabilizing decomposable organic matter under aerobic conditions (Sawyer and McCarty, 1967). It is expressed as either the mass of oxygen utilized per volume of waste present or the mass of oxygen per mass of waste present. The BOD test is an empirical test using standardized laboratory procedures. The standard time for conducting the test is five days; many tests are reported as BOD_5. The ultimate value of BOD as time approaches infinity (BOD_u) is determined by extrapolating laboratory results to longer times.

Chemical oxygen demand (COD) is an indirect measure of the concentration of oxidizable material present in an organic waste. The COD test allows measurement of the total quantity of oxygen required for oxidation of organic matter to carbon dioxide and water, but it does not differentiate between biologically oxidizable and biologically inert organic matter (Sawyer and McCarty, 1967). COD is expressed as the mass of oxygen utilized either per volume or per mass of the waste present. COD values are several times larger than BOD values. COD is also determined by means of a standardized laboratory procedure. Since considerable difficulty has been experienced in obtaining reproducible BOD tests on animal manures, COD tests appear to be easier to obtain and more useful for work with animal manures.

The total solids (TS) content of a waste is the residue after water is evaporated from the sample by heating to 103°C. If a waste is liquid, the total solids may be divided into *suspended solids* and *dissolved solids*. Suspended solids are the solids in suspension that could be removed by ultrafiltration. Dissolved solids are those that are dissolved in the sample and precipitate out during drying.

Volatile solids (VS) are the solids driven off as a gas when the total solids are incinerated at 600°C for one hour. The solids remaining after incineration are referred to as *fixed solids*. The volatile solids content of a waste is a measure of the amount of decomposable organic matter in the waste. Solids content can be further classified into *volatile suspended* and *volatile dissolved solids*, or *fixed suspended* and *fixed dissolved solids*.

Biodegradability of Organic Matter

Organic matter must be degraded by anaerobic bacteria (acid- and methane-formers) to produce methane. The biodegradability of organic matter must be considered to predict methane yield. The dry matter content of biomass feedstocks can be partitioned into ash and volatile solid fractions. Volatile solids are composed of cellulose, hemicellulose, lignin, crude protein, and other materials (Table 3.1).

The biodegradable fraction of corn stover, grass, and wheat straw was predicted using the following equation, developed from experimental data by Chandler et al. (1980):

$$B = 0.830 - 0.028X \tag{1}$$

where B is the biodegradable fraction of the total volatile solids (TVS) and X is the lignin content (measured by sulfuric acid method), percentage of TVS. Average biodegradabilities at 35°C predicted by this model were as follows: corn, 0.65; grass, 0.60; wheat straw, 0.56; and dairy-cow manure digester effluent, 0.54 (Jewell et al., 1982).

Anaerobic Fermentation

Anaerobic fermentation is a biological process in which organic matter is decomposed without atmospheric oxygen, to yield water, carbon dioxide, and methane. The phenomenon occurs naturally when organic material remains without oxygen under conditions conducive to

TABLE 3.1 Fiber Composition of Anaerobic Digester Substrate

Substrate	Total Volatile Solids (% total solids)	% Total Volatile Solids			
		Cellulose	Hemi-cellulose	Lignin	Crude Protein
Corn stover	92.6	36.9	32.0	6.5	5.8
Wheat straw	95.7	42.7	29.0	9.6	7.5
Grass	95.3	41.5	29.7	8.3	7.2
Dairy cow manure, digester effluent	89.3	24.3	25.4	10.2	18.0

Source: Jewell et al. (1982).

microbial growth. Such conditions occur in many natural environments, ranging from pond sediments to the gastrointestinal tract of animals.

Microbiology

According to Bryant (1979), methane formation occurs in three distinct stages. First, fermentative bacteria hydrolyze polysaccharides, lipids, and proteins and ferment the products to short-chain organic acids, alcohols, hydrogen (H_2), carbon dioxide (CO_2), and ammonia (NH_3). Second, acetogenic bacteria oxidize the products of the fermentative bacteria to acetate, CO_2, and H_2. Third, methanogenic bacteria use H_2 to reduce CO_2 and formate to form methane (CH_4). Some species also cleave acetate to produce CO_2 and CH_4.

The chemical composition of organic wastes consists of carbohydrates, proteins, and lipids. Carbohydrates exist primarily as polysaccharides such as starch and cellulose. These polysaccharides are hydrolyzed by extracellular enzymes to monosaccharides made up primarily of glucose with some fructose and mannose. The monosaccharides are further broken down to organic acids and alcohols. The major organic acids produced by the acid-forming bacteria are acetic and propionic (or butyric), of which acetic is the most important. Studies by McCarty (1964a) indicate that 72 percent of the methane produced is formed from acetic acid (CH_3COOH) and 13 percent from propionic acid (CH_3CH_2COOH) or butyric acid (C_3H_7COOH), as shown in

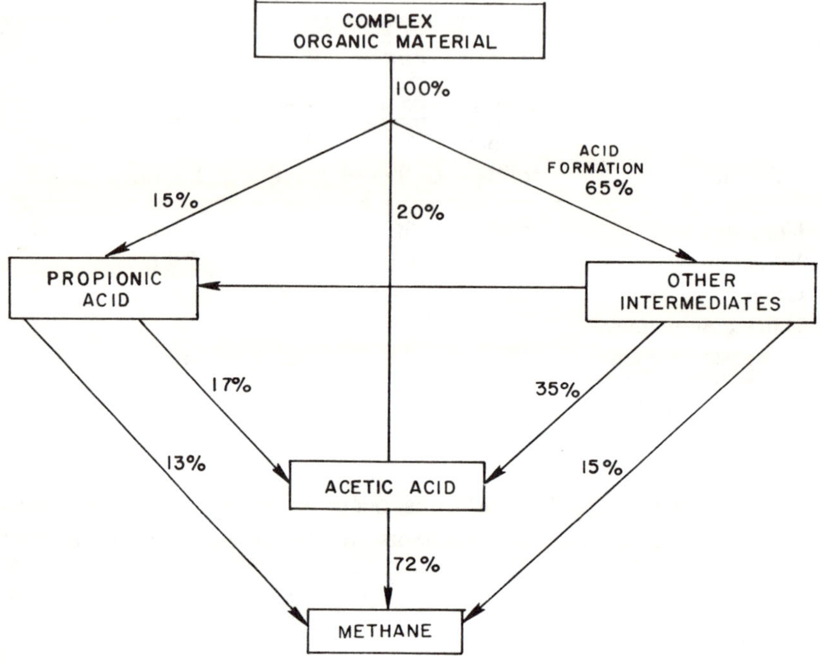

FIG. 3.2. Methane formation from a complex organic material; percentages refer to the conversion of chemical oxygen demand, or COD (from McCarty, 1964a).

Figure 3.2. The remaining 15 percent is formed from H_2 and CO_2, reduction of methanol, and degradation of other intermediate compounds.

Barker (1965) postulated the three methane forming routes of interest as follows:

$$4CH_3CH_2COOH + 2H_2O \rightarrow 4CH_3COOH + CO_2 + 3CH_4$$

$$CH_3COOH \rightarrow CH_4 + CO_2$$

$$CO_2 + 4H_2 \rightarrow CH_4 + 2H_2O$$

Jeris and McCarty (1965) showed that long-chain fatty acids can be fermented via β-oxidation. This requires removal of a two-carbon fragment from the end distant from the carboxyl and the conversion of this fragment to acetate, which is then fermented to CO_2 and CH_4 by methane formers. Since H_2 is produced during the β-oxidation, it may be combined with CO_2 to form more methane.

Environmental Factors

Because anaerobic fermentation is a complicated biological process, it is very sensitive to several environmental factors: temperature, pH, organic loading rates, influent-solids concentration, nutrient availability, and toxic substances.

Temperature, an extremely important factor, controls the microbial growth rates during the fermentation process. Three thermal zones of microbial activity have been identified: psychrophilic (below 28°C), mesophilic (28–42°C), and thermophilic (above 42°). Methane production in the psychrophilic zone is small and will not be discussed further; the mesophilic and thermophilic are the zones of primary interest for commercial methane production (Figure 3.3).

The methane production bacteria are sensitive to small changes in pH. During normal methane operations, the pH is maintained naturally by the bicarbonate buffer system resulting from the CO_2 produced during anaerobic fermentation. If the equilibrium balance is altered and too much organic acid is produced, then the pH of the system may drop below 6.6 and the methane forming bacteria die off rapidly without producing methane. The optimum pH range is between 7.0 and 7.2.

The organic loading rate is the amount of waste added per day per unit volume of digester. The influent-solids concentration is the ratio of the mass of dry solids to the total mass of the waste (dry solids plus water), expressed as a percentage. Available data indicates that anaerobic digestion has been successful with organic loading rates of 1.6 to 3.2 kg of volatile solids per day per m^3 of digester and influent-solids concentration of 6 to 12 percent.

As with other life forms, a balanced diet must be maintained for the microbes. Microbial cells require carbon, nitrogen, phosphorus, and micronutrients for adequate growth. Ratios of 100:2:0.5 (available organic carbon to nitrogen to phosphorus) are necessary to maintain nutritionally balanced cell growth.

Many organic and inorganic substances may be toxic to the microbes in the digester if concentrations are excessive. The toxic substances of principal concern are heavy metals, alkaline metals (Mg, Ca, Na, K), ammonia, and soluble sulfides.

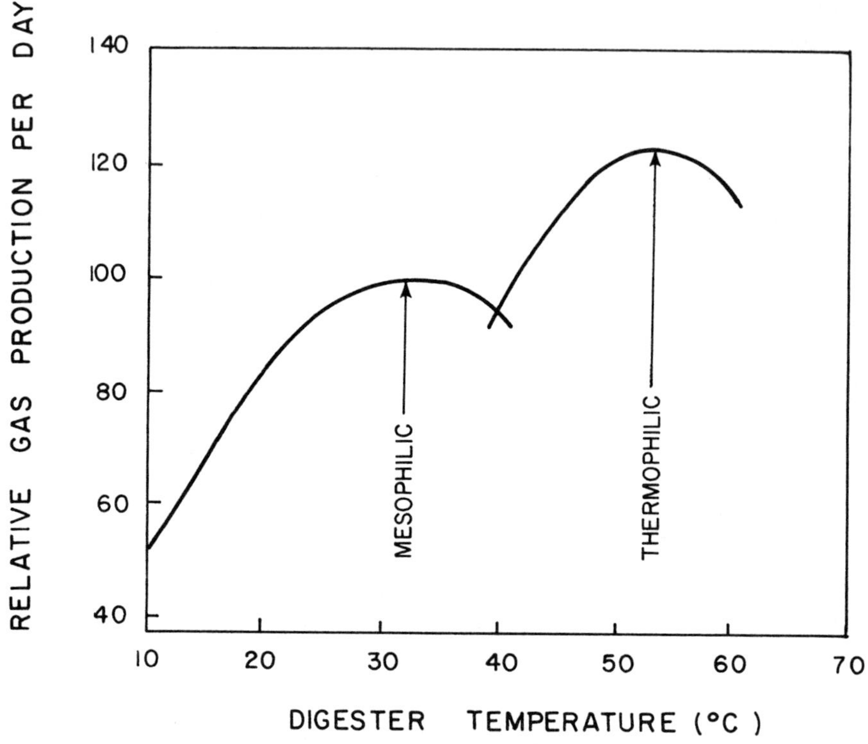

FIG. 3.3. Effect of temperature on gas production rate (from Roediger, 1967).

Fermentation Kinetics

The development of kinetic models for methane production from animal wastes has been a slow process. Three types of kinetic models are discussed by Hill (1982a): Monod kinetic models, first-order kinetic models, and Contois kinetic models. The Monod models possess the ability to predict, often with high accuracy, process variables and points of failure. But these models require that 15 or more kinetic parameters be evaluated for each set of operating conditions and each waste type. Some of these parameters cannot be evaluated for complex wastes.

The first-order kinetic models simplify the input parameters but do not predict optimum operating conditions or process failure. Contois (1959) developed a kinetic model of bacterial growth that possesses advantages of both Monod and first-order kinetic models. Chen and Hashimoto (1979a) modified and adapted the Contois model to animal waste digestion as follows:

$$\gamma_v = \frac{B_o S_o}{\theta}\left(1 - \frac{K}{\theta\mu_m - 1 + K}\right) \tag{2}$$

where γ_v = volumetric methane productivity (L CH_4 per L digester per day),
B_o = ultimate methane yield (L CH_4 per g VS added as $\theta \rightarrow \infty$),
S_o = influent–volatile solids concentration (g VS per L waste),
θ = detention time (days),
μ_m = maximum specific growth rate of microbes in the digester (day^{-1}), and
K = kinetic parameter (dimensionless).

The derivation of equation (2) did not include a death term for the microbes. Hill (1982b) considered the death of the microbes to be important and modified equation (2) to give

$$\gamma_v = \frac{B_o S_o}{\theta}\left(1 - \frac{K(1 + K_D\theta)}{\mu_m\theta - (1 + K_D\theta)(1 - K)}\right) \tag{3}$$

where K_D is the specific death rate of microbes (day^{-1}), and all other terms are as previously described.

The ultimate methane yield (B_o) is an expression of the biodegradability of the waste. The ration, age of the manure, animal species, and amount of foreign material (dirt and bedding) greatly affect the value of B_o. McCarty (1964b) reported a theoretical methane productivity (γ_t) of 0.5 L CH_4 per g of VS destroyed. The ultimate VS destruction (V_u) is expressed as

$$V_u(\%) = \frac{100 \cdot B_o}{\gamma_t} \tag{4}$$

Hill (1982a) presented selected values of B_o and V_u for the four major animal species of beef, dairy, swine, and poultry. These data are shown in Table 3.2.

TABLE 3.2 Ultimate Methane Yield (B_o), from Manure and Predicted Ultimate Volatile Solids Destruction (V_u)

Animal Species	B_o (L CH_4/g VS)	V_u (%)
Beef cattle	0.35	70
Dairy cattle	0.20	40
Swine	0.45	90
Poultry	0.39	78

Source: Hill (1982a).

The maximum specific microbial growth rate (μ_m) appears to be the same for all waste types (Hill, 1982a). Temperature is the only variable that appears to influence μ_m. Suggested values of μ_m as a function of temperature are shown in Figure 3.4.

The value of the kinetic parameter (K) is a function of the animal type and the influent–volatile solids concentration (S_o) as shown in Figure 3.5 (Hill, 1982a). When B_o, S_o, θ, and μ_m are constant, equations (2) and (3) show that methane production (γ_v) decreases as K increases. Thus, maximum methane production will be decreased for influent–volatile solids concentrations above 80, 80, 50, and 40 g VS/L for dairy cattle, beef cattle, swine, and poultry, respectively.

The methane production (γ_v) is decreased as the specific death rate K_D increases; see equation (3). Hill (1982b) suggested that the maximum value of K_D is 0.1 μ_m, and used this value in his analysis.

Total methane production (P) in L/day is given by

$$P = \gamma_v V \tag{5}$$

where V is the volume of digester (L).

Combining equations (2) and (5) yields

$$P = \frac{B_o S_o V}{\theta}\left(1 - \frac{K}{\theta\mu_m - 1 + K}\right) \tag{6}$$

For a given loading rate ($S_o \cdot V/\theta$), the maximum methane production will occur when the maximum VS destroyed (V_m, g/day) occurs. This will occur at steady state. Under these conditions,

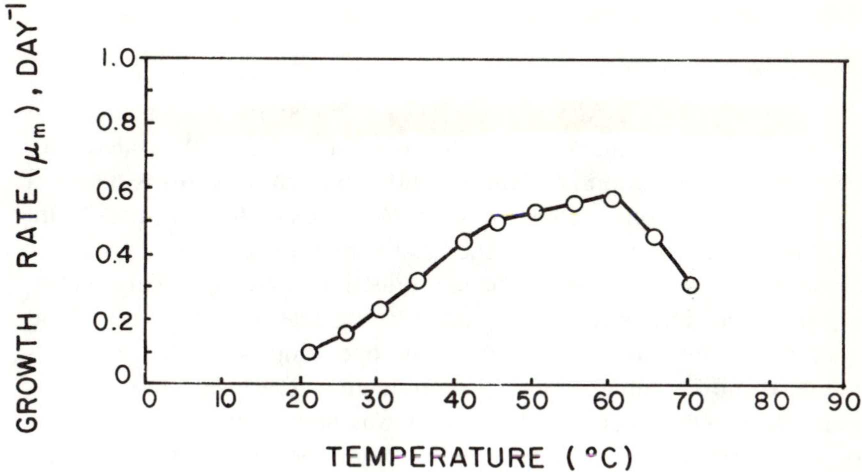

FIG. 3.4. Maximum specific microbial growth rate (μ_m) curve (after Hill, 1982a).

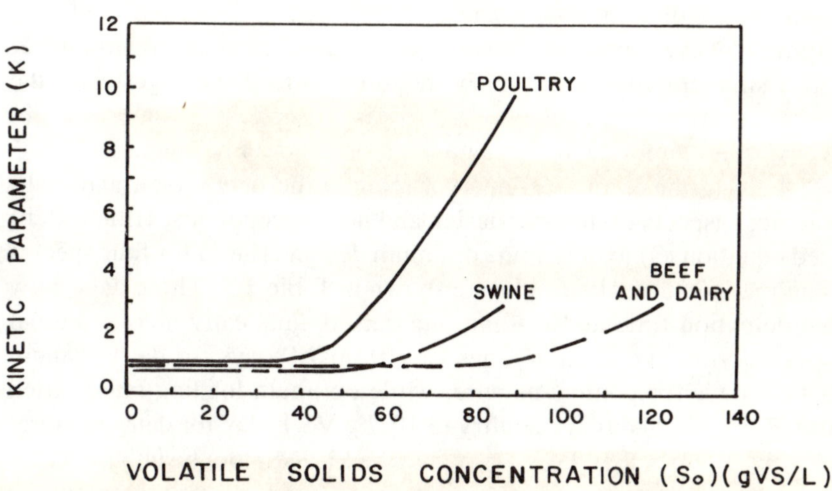

FIG. 3.5. Kinetic parameter (K) as a function of influent–volatile solids concentration (S_o) and waste type (after Hill, 1982a).

$$\frac{K}{\theta\mu_m - 1 + K} = 0 \tag{7}$$

Two different design approaches can now be undertaken. The first approach involves maximizing the volumetric methane productivity (γ_v) given by equation (2). The second approach maximizes the total methane production given by equation (6). The condition of maximum total methane production is defined as the operating point where, over a one-day period of time, methane production is greatest. This operating point, as determined by organic loading rate, detention time, and operating temperature, is not the same operating point that produces the maximum volumetric methane production. For instance, the maximum volumetric methane production was shown by Hill (1982a) to occur at temperatures of about 60°C, while the maximum total daily methane production occurred at about 45°C. Further, Hill (1982a) showed that over a given period of time the digester designed for maximum total methane production will produce 1.43, 1.74, 1.50 and 1.60 times the methane produced by a digester designed for maximum volumetric methane productivity for dairy cattle, poultry, swine, and beef cattle, respectively.

Total capital costs of the plant designed for maximum methane production will be greater because of the larger digester required. Net return, however, will be greater because more methane is produced, and a smaller proportion will be required to heat the digester to the operating temperature of 60°C. A detailed economic analysis is required to determine which design approach should be used.

Letting the term "optimized" refer to the operational and technological aspects of digester design and not to economics, Hill (1982b) used equation (3) to determine optimum design criteria for four species of animal manure. His results are shown in Table 3.3. These data show that detention time and loading rate vary dramatically among animal types. Optimum detention times vary from 7.9 days for dairy manure to 14.8 days for poultry manure, while optimum loading rates varied from 2.5 g VS/L-day for poultry to 10.7 g VS/L-day for dairy manure. This simply says that dairy waste may be loaded more heavily and processed for a shorter time than poultry manure to reclaim the maximum amount of methane.

The optimal design parameters obtained by Hill (1982b) to produce a maximum volumetric methane productivity (γ_{vm}) are shown in Table

TABLE 3.3 Optimized Design Parameters for Anaerobic Digestion of Manure for Maximum Total Methane (P)

Parameter	Units	Animal Species			
		Swine	Beef	Dairy	Poultry
Detention time, θ	days	11.7	10.3	7.9	14.8
Temperature, T	°C	38	40	39	32
Volumetric loading rate, S_o/θ	g VS/L·day	4.8	8.1	10.7	2.5
Influent–total solids concentration, TS	%	7.0	9.7	10.3	5.3
Volatile solids reduction, V_m	%	65.7	48.0	24.5	45.5
Volumetric methane productivity, γ_v	L CH_4/L·day	1.6	1.9	1.3	0.6

Source: Hill (1982b).

3.4. Note that the detention time for all wastes is in the 3.6- to 4.1-day range. This shows that methane productivity reaches a plateau very rapidly and suggests that thermophilic digestion converts only the more readily available carbon into methane, leaving a great deal of potential methane behind in the less readily degradable fraction.

Anaerobic Digester Types

A variety of anaerobic digester design concepts were described by Stafford, Hawkes, and Horton (1980) and Ashare and Wilson (1979). Anaerobic digesters can be classified according to loading frequency, temperature, degree of mixing, configuration, and construction materials. The simplest digesters are batch loaded. With the biomass loaded, the vessel is sealed and left until digestion is completed (Stafford, Hawkes, and Horton, 1980).

Most anaerobic digesters are loaded daily, with fresh feedstock displacing a like amount of digested slurry. With daily loading, equilibrium conditions are maintained, and gas production rate and methane concentration is relatively constant over many months. Some field experiments have been conducted with batch-loaded digesters in which peak yields of methane are reached a few weeks after start-up, after

TABLE 3.4 Optimized Design Parameters for Anaerobic Digestion of Manure for Maximum Volumetric Methane Productivity (γ_{vm})

Parameter	Units	Animal Species			
		Swine	Beef	Dairy	Poultry
Detention time, θ	days	3.6	3.8	3.8	4.1
Temperature, T	°C	60	60	60	60
VS concentration	g VS/L	55	85	85	45
Max. volumetric methane productivity, $\gamma_{v,\ max}$	L CH_4/L·day	3.3	3.5	2.0	1.7
Influent–total solids concentration, TS	%	6.9	9.9	10.3	6.4

Source: Hill (1982b).

which yields taper off. Batch-loaded digesters are usually staggered in start-up sequence in an attempt to achieve high and more uniform methane production.

Anaerobic bacteria perform most efficiently in terms of gas production at temperatures of 35°C (mesophilic) and 57°C (thermophilic). Thermophilic digesters provide peak methane yields and can be one-half to two-thirds smaller than mesophilic digesters. However, they suffer higher heat losses and are more subject to bacterial upset.

Digester contents must be mixed to distribute temperatures and renew bacterial contact with feedstock and to release gases from decomposition. High-rate digesters incorporate mechanical stirring or agitation with compressed gas. However, it is desirable to keep mixing to a minimum to conserve energy. One researcher arrived at a mixing frequency of 10 to 12 minutes per hour (Fulton, 1979, 1981). An exceptional case is the so-called plug-flow digester, in which no mechanical mixing is utilized. These digesters, usually long and narrow, apparently work best with fresh dairy manure at 12 percent total solids concentration, which does not stratify in the digester (Hayes, 1979).

Other types of anaerobic digester configurations are included in the following list (Stafford, Hawkes, and Horton, 1980).

1. The *attached-film anaerobic filter* employs a packed bed of inert material, such as plastic or a fabric, which retains a film of methanogenic bacteria and is suitable for operation at short retention time

(two to six days) on moderate strength wastewaters containing, for example, 10,000–30,000 mg/L BOD_5 (Norstedt and Thomas, 1983; Jones et al., 1981).

2. The *anaerobic contact process* uses an effluent sludge settling tank or separator to capture and recycle sludge into the influent stream; and it could be operated at short retention time on moderate strength wastewaters.

3. *Multistage digesters* are two or three-stage anaerobic digesters in series, designed to optimize conditions for fast-growing acid-forming and more sluggish methane-forming bacteria in successive stages.

4. The *expanded bed reactor* is an upflow digester unit in which fine inert bed material, such as sand, captures solids and retains bacteria for effluent digestion in a short retention time.

Anaerobic digesters are usually built of concrete and/or steel wrapped with heavy insulation or buried to retain heat. Most designs have been patterned after the rigid-wall, complete-mixing digesters used for sewage sludge stabilization for many years, where neither cost nor maximum gas production are serious concerns.

For methane production to be practical in agriculture, the cheapest digester that will reliably produce biogas at acceptable levels is a necessity. Rigid-wall, aboveground mesophilic anaerobic digesters, complete with slurry and gas-handling systems, may be prohibitively expensive. Some universities and companies have developed digesters installed in trenches lined with concrete or flexible membrane materials (Hayes, 1979; Kemp, 1978; Fulton, 1980). These digesters reduce total system cost substantially and offer potential for making anaerobic digestion economically attractive (Figure 3.6).

Operational Modes

Four basic anaerobic digestion designs are currently used to digest agricultural wastes: (1) continuous-flow, (2) batch-flow, (3) plug-flow, and (4) continuously expanding digesters.

In the conventional continuous-flow process, a set volume of waste is added daily and an equal volume of effluent removed, keeping the digester volume constant. Different modes of mixing the digester contents have been used, all the way from continuous mixing to no mixing.

The batch-load process uses two digesters operating in parallel.

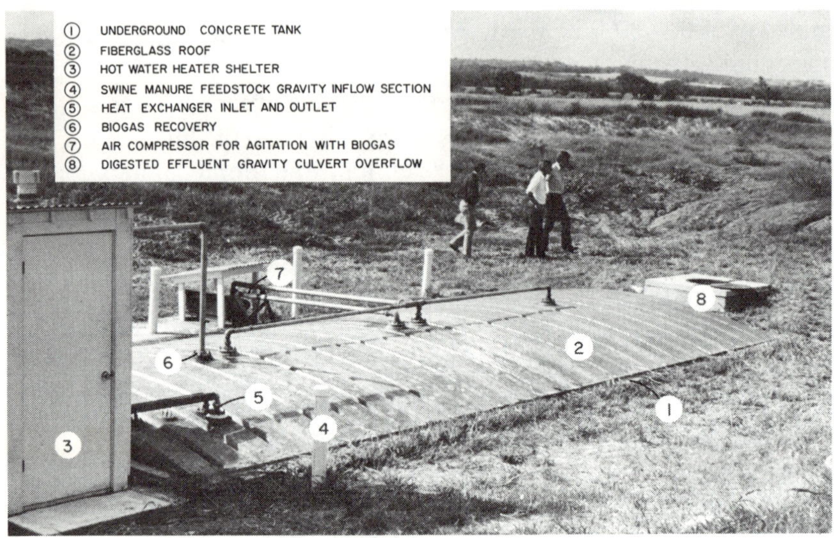

FIG. 3.6. Anaerobic digester on a commercial Texas swine farm: a concrete-lined underground tank with heat exchanger coils and rigid fiberglass roof for gas collection (from Malish, Raley, and Berdoll, 1982).

The first is loaded with a given daily volume until it is filled. When the first digester is full, the second is then loaded with the same daily volume. While the second digester is being filled, the first digester is allowed to ferment without additional loading. At the end of the loading period for the second digester, the first is emptied, and the process is repeated. Mixing may be either continuous or intermittent.

In plug flow, influent is introduced at one end of a longitudinal digester and effluent withdrawn from the other. No internal mixing is used in this process.

The continuously expanding digester (CED) is a recently introduced process (Hill, Young, and Nordstedt, 1981) in which a daily loading is made, but no effluent is removed. Thus, the digester volume gradually expands over a cycle of some 90 to 180 days; then the effluent is removed and applied to land. This process may be termed a "semi-batch" mode. Its advantage is that the operator does not have to dispose of effluent on a daily basis. Its disadvantage is that it requires a large volume, which may be expensive.

Batch-Load Digester

Results from laboratory work conducted by Jewell et al. (1976) indicated that a batch-load method of digester operation might be more efficient than the continous-flow method. Egg (1979) evaluated the batch-load method at Texas A&M University, using dairy manure under both laboratory and field conditions.

Laboratory Studies. Egg (1979) conducted a laboratory study of batch-load digestion using three 2-liter digesters (Figure 3.7). Influent–volatile solids concentration was the variable of interest, while other parameters were held constant. All digesters were operated with a 20-day batch time, and temperature was equal to 32°C ± 2°C. Fresh dairy feces with little or no urine was used in the digesters. In the operational mode, 20 percent of the digester volume (0.4L) was maintained as seed. Manure slurry was added to the digester for ten days. Digesters were mixed twice daily. Fermentation was continued for an additional ten days with no slurry added. Afterward, 1.6 L of effluent was removed from the digester. The cycle was repeated through a total of eight runs at different volatile solids loading rates.

The results of the laboratory studies are summarized in Table 3.5, in which the influent characteristics are average daily loadings over a ten-day period. Gas samples were analyzed, using a gas chromatograph, to obtain the percentage of CH_4 and CO_2. The CH_4 fraction varied from 52 to 81 percent, averaging 65 percent. There was a strong tendency for the low CH_4 percentages to occur during the first ten days, after which they gradually increased to the high values of 78 to 81 percent during days 11 through 20.

(In the first experiment, the volatile solids–influent concentration of 96.5 g VS/L was evidently too high, because pH declined to 5.4 and essentially no methane was produced. No lime was added to assist the digester during start-up. Results from that experiment are not included in Table 3.5.)

The specific methane productivity (γ_s) appeared to be relatively constant for all runs and averaged 0.45 L CH_4/g VS removed, which compares favorably with the theoretical value of 0.5 L CH_4/g VS removed, as was reported by McCarty (1964a). This indicates that the batch-load digester was operating very close to the theoretical maximum.

The volatile solids reduction (V_m) varied from 32 to 50 percent,

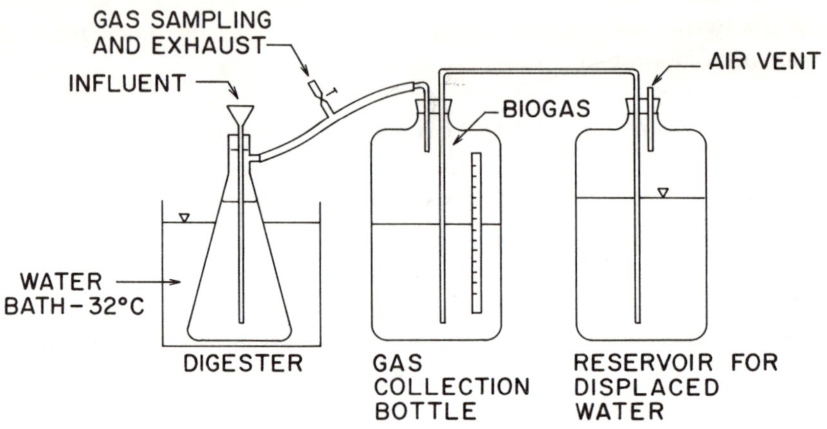

FIG. 3.7. Schematic of laboratory digester (from Egg, 1979).

and was apparently not related to the loading rate. The volatile solids removed averaged 43 percent for all runs, which is considered to be very good.

By comparison, Hill, Young, and Nordstedt (1981) stated that volatile solids removal for dairy waste with continuous-flow digesters has been reported in the literature to range from 21 to 44 percent. They obtained a 61 percent volatile solids removal with dairy waste, using a continuously expanding digester.

Data from the laboratory-scale batch digester (Table 3.5) show that the volumetric methane productivity (γ_v) was strongly related to the volatile solids inflow rate, as shown in Figure 3.8. Data and theoretical considerations presented by Hashimoto, Chen, and Prior (1979b) suggest that γ_v peaks at a VS inflow concentration of 80 g VS/L (about 8 percent volatile solids) for continuously loaded thermophilic digesters. Thus, there is evidence to suggest that γ_v may have been approaching a peak at 0.45 L CH_4/L digester-day and a VS inflow concentration of 60 g VS/L (Figure 3.8).

TABLE 3.5 Chemical Characteristics, Batch-Load Laboratory Digester

Constituent	Experiment Numbers 2–8 Range	Average	Standard Deviation
Influent characteristics			
TS (g/L)	34.6–81.0	59.0	17.8
VS (g/L)	24.3–65.1	44.8	14.0
pH	5.8–6.4	6.0	0.2
Effluent characteristics			
TS (g/L)	19.3–48.0	35.2	10.4
VS (g/L)	12.6–37.2	25.7	8.8
pH	6.8–7.4	7.2	0.2
Digester performance			
Total biogas (L)	12.77–26.22	20.4	5.5
Total methane[(a)](L)	8.30–17.97	13.3	3.5
Biogas production (L/g VS removed)	0.59–0.82	0.69	0.07
Specific methane productivity,[(b)] γ_s (L CH_4/g VS removed)	0.38–0.53	0.45	0.05
Volatile solids reduction, V_m (%)	32.4–50.2	42.8	6.3
Volumetric methane productivity,[(c)] γ_v (L CH_4/L digester-day)	0.21–0.45	0.33	0.09

Source: Egg (1979).
(a) Methane calculated as 65% of total biogas production.
(b) γ_s calculated as (L CH_4)/(1.6 VS_{in} − 1.6 VS_{out}).
(c) γ_v calculated as (L CH_4)/(20-day · 2 liters).

It is difficult to compare values of γ_v from various literature sources because no standardized format for reporting γ_v has been devised for batch digesters. In Table 3.5, the γ_v data were calculated by dividing the total gas production by the total volume of the digester (2 liters) and the batch detention time (20 days). Since part of the digester volume (0.4 liter) is inactive in the batch digester and actual loading is done for only the first ten days with no material added on days 11 through 20, a comparison of γ_v for batch digesters with continuous-flow digesters is difficult. However, it would appear from the literature

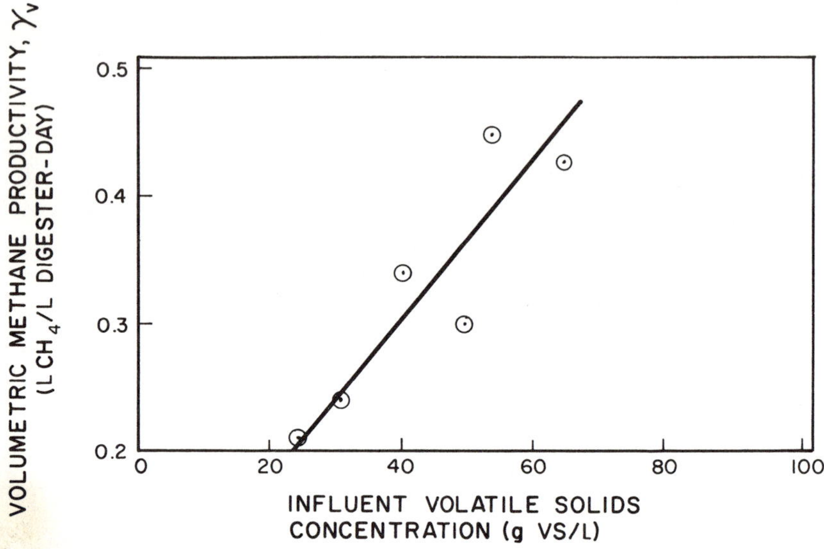

FIG. 3.8. Volumetric methane productivity (γ_v) as a function of the influent-volatile solids concentration (adapted from data in Egg, 1979).

that values of γ_v for continuous-flow digesters using dairy manure have varied from 0.56 to 1.41 L CH_4/L digester-day, two to three times larger than the values reported by Egg (1979) for batch-load digesters. This means that larger digesters (more capital investment) will be required for the batch process than for continuous flow.

Field-Scale Batch Digester. A schematic of the field-scale batch digester used by Egg (1979) is shown in Figure 3.9. The digester was a 3.2 m^3 fiberglass tank with a fixed cover buried in the soil for insulation purposes. Temperature was controlled at 32°C ± 2°C by pumping hot water through a heat exchange coil inside the digester. The operating volume of the digester was set at 2 m^3. The seed volume was 0.4 m^3, and the influent flow rate was 0.16 m^3 per day for ten days. The digester received no influent on days 11 through 20. The digester was mixed via a recirculating pump for one hour each day. Three 20-day batch cycles were made.

Results from the field-scale digester (Table 3.6) were compatible with the results from the laboratory experiments (Table 3.5). The field-

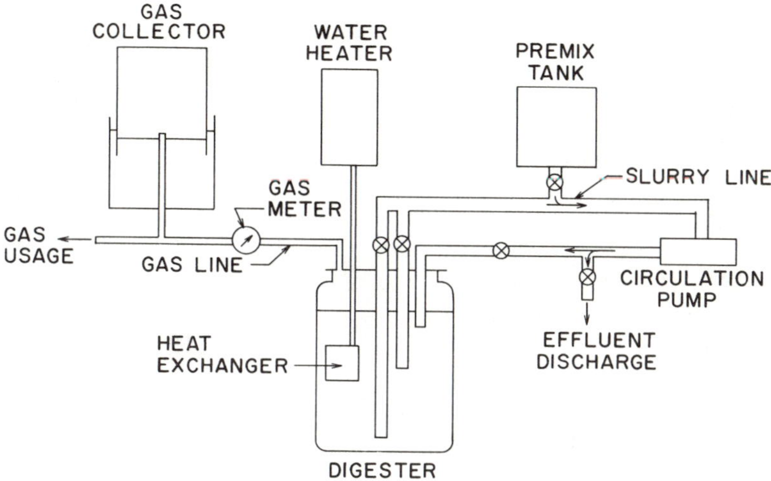

FIG. 3.9. Schematic of field-scale digester (from Egg, 1979).

scale batch digester had good volatile solids reduction (48 percent) and specific methane productivity (0.55 L CH_4/g VS removed). The volumetric productivity (γ_v) is one-half to one-third of that for continuous-flow digesters.

Continuously Expanding Digesters

Hill, Young, and Nordstedt (1981) found that CEDs could be used for periods of 90 to 200 days without reduction of digester volume. They could also operate at 21°C just as effectively as at 35°C, thus eliminating the need to heat the digester. As shown in Table 3.7, the CED has γ_s values very close to the theoretical γ_t of 0.5 L CH_4/g VS removed. Also, from Table 3.7, V_m of 50 to 60 percent for CEDs is much superior to the 20 to 40 percent obtained with continuous-flow digesters. The main disadvantage of the CED is that it requires approximately four times more digester volume than continuous-flow digesters. The advantages are that (1) daily withdrawals of effluent are unnecessary; (2) the digester is relatively insensitive to intermittent or shock loadings with manure; and (3) a higher degree of waste treatment is obtained, partly because of the long retention times.

TABLE 3.6 Chemical Characteristics, Batch-Load Field-Scale Digester with Dairy Manure

Constituent	Avg. ± Std. Dev., 3 Experiments
Influent characteristics	
TS (g/L)	52.9 ± 8.7
VS (g/L)	38.8 ± 9.4
pH	6.6 ± 0.3
Effluent characteristics	
TS (g/L)	29.5 ± 0.1
VS (g/L)	19.8 ± 1.5
pH	7.3 ± 0.1
Digester performance	
Total biogas (L)	23,190 ± 1,290
Total methane[(a)](L)	15,070 ± 840
Biogas production (L/g VS removed)	0.84 ± 0.36
Specific methane productivity,[(b)] γ_s (L CH_4/g VS removed)	0.55 ± 0.23
Volatile solids reduction, V_m (%)	48.1 ± 8.5
Volumetric methane productivity,[(c)] γ_v (L CH_4/L digester-day)	0.38 ± 0.02

Source: Egg (1979).
(a) Methane calculated as 65% of total biogas production.
(b) γ_s calculated as (L CH_4)/(1,600 VS_{in} − 1,600 VS_{out}).
(c) γ_v calculated as (L CH_4)/(20-day · 2,000 liters).

TABLE 3.7 Specific Methane Productivity of CED and Volatile Solids Removed

Temperature °C	Specific Methane Productivity, γ_s (L CH_4/g VS removed)	Volatile Solids Removed, V_m (%)
21.1	0.477	51.2
35.0	0.495	61.1

Source: Hill, Young, and Nordstedt (1981).

Continuously Loaded Digesters

Most anaerobic digesters receive daily loadings of feedstock to approximate a continuously loaded condition. Digested effluent displaced by incoming feedstock overflows by gravity to a processing area, lagoon, or storage pond.

Fulton (1978, 1980) installed two continuously loaded anaerobic digesters. The first was an aboveground, cylindrical steel storage tank insulated and converted for use as a 26.5 m^3 mesophilic digester (Fulton, 1978). Fresh chicken manure from an average of 2,700 caged laying hens (2,200–3,500 bird range) was scooped into a receiving pit and mixed with enough water to achieve an 8 to 9 percent total solids concentration. Slurry was loaded into the digester daily. The digester was operated with a 16-day detention time (14- to 22-day range) and a temperature of 35°C. Biogas production was 25 m^3 (60 percent methane) per day. An average of one-fourth of the gas produced was needed to maintain the digester temperature at 35°C; the remainder was available for other uses. Digested effluent was discharged by gravity from 0.25 m below the liquid surface, and sludge was pumped from the bottom at weekly intervals to maintain adequate mixing. The volatile solids destruction rate was 48 percent (Fulton, 1978).

Fulton (1980) determined that the first five months of start-up are critical to the successful and prolonged operation of an anaerobic digester under typical farm conditions. Maintaining the correct temperature, loading rate, loading schedule, and mixing rate were important during start-up to allow acclimatization of bacteria to the substrate and development of a proper buffering capacity. After start-up, temperature control was the most important maintenance variable. When bacteria had become acclimated, manure loading was halted for 2½ weeks, and 95 percent of the steady-state biogas yield was observed after only three days (Fulton, 1978).

Fulton (1980) installed a second anaerobic digester based on 3½ years of research with the first unit. This mesophilic unit consisted of a rectangular concrete-lined buried pit with an arched rigid-fiberglass roof (see, for example, Figure 3.6). The digester had a liquid volume of 113 m^3 and could handle manure from 16,000 caged layers, 100 dairy cows, or 40 sows (farrow-to-finish operation). Digester dimensions were 4.2 m wide, 9.1 m long, and 3.0 m deep. The liquid level was maintained by a weir of large-diameter pipe, which allowed volume control and maintained anaerobic conditions. The digester contents were mixed by recirculating biogas with a lobe compressor; temperature was maintained at 35°C by circulating water through a heat exchanger. The digester was loaded daily with about 7.0 m^3 of manure slurry (9 to 10 percent total solids concentration) to provide a 16-day retention time. Gas pressure inside the digester was maintained at 50 mm water. This digester was designed for minimum labor, management, and maintenance.

Batch-Loaded Leachate Recycle System

Batch-loaded fermentation systems for methane production from crop residues containing 15 to 40 percent total solids content were conceived and constructed by Jewell et al. (1982). The so-called dry fermentation system was investigated in laboratory-scale (1L), pilot-scale (5 m^3), and field-scale (110 m^3) digesters. Feedstocks tested were grass, wheat straw, and corn stover; temperatures were 35 and 55°C.

The mesophilic field-scale dry-fermentation unit was designed to collect leachate at the bottom, then to pump and spray it over the top of the wheat straw feedstock to redistribute moisture, nutrients, and bacteria without mixing the high-solids digester contents (Jewell et al., 1982). At 26 percent total solids content, insufficient leachate was produced to allow leachate recycling; after water was added to lower the total solids concentration of digester contents to 20 to 21 percent, however, leachate recycling was used intermittently. The inoculum-to-feedstock ratio was initially 0.30. The biogas and methane production rates realized over a 100-day batch fermentation period were 0.25 and 0.12 m^3/m^3/day, respectively. Total volatile solids destruction was 26 percent after 100 days and 48 percent after 280 days, which represented 89 percent conversion of biodegradable volatile solids. Energy requirements for temperature maintenance at 35°C were less than 15 percent. The decay coefficient for both the 110-m^3 field-scale unit with wheat straw and the 5-m^3 pilot-scale units with grass was 0.0175/day.

Future Work

Research with high-solids anaerobic digesters has begun at Texas A&M University, with sweet sorghum plants used as the feedstock for methane production (Hiler, 1982). Laboratory digesters (2L) will be installed to examine the effects of solids content, inoculum-to-feedstock ratios, and varying reactor designs on methane yield at 35°C. Four types of pilot digesters (18 m^3 each) will be comparatively tested: (1) batch-loaded, (2) batch-loaded with leachate recycle, (3) batch-loaded with leachate recycle and upflow packed-bed methanogenic stage, and (4) conventional stirred tank. Solids contents, inoculum levels, and detention times will be selected from the laboratory digester results. Other major components of the research project are sweet sorghum breeding, production, harvesting, and storage for methane production.

Digester Design

The design of anaerobic digesters is based on several parameters. Feedstock characteristics of major importance include ash content, carbon content, carbon-nitrogen ratio, ultimate biodegradability, ultimate methane yield, and specific gas production rate.

A second design factor is detention time (θ). The longer the organic solids residence time, the larger the digester needs to be. Gas production occurs rapidly at first but begins to level off after 10 to 20 days with most types of feedstocks. Correct design involves selecting the best trade-off between gas yield and digester construction cost. For mesophilic systems, detention times of 12 to 20 days are usually used. A highly digestible feedstock, such as stillage from ethyl alcohol production, may only require 2 days $\leqslant \theta \leqslant$ 5 days, while resistant materials may require θ values of 30 days or more.

A third design factor is feedstock solids concentration, which dictates digester size, performance, and cost. To minimize digester volume, solids concentrations should be as high as possible without causing mixing problems or overloading. Desirable concentrations of total solids are 8 to 20 percent, with most completely mixed systems operating at 8 to 12 percent total solids. Jewell et al. (1982) designed and operated mesophilic anaerobic digesters with corn stover, grass, and wheat straw feedstock at total solids concentrations of 15 to 40 percent (so-called dry-fermentation mode). Hashimoto, Chen, and Prior (1979b) determined that the optimum volatile solids (VS) concentration for thermophilic digester influent is 7 percent with beef cattle manure.

A fourth design factor is loading rate, which is usually specified in terms of the weight of volatile solids added to the digester daily per unit of digester volume. Typical volatile solids loading rates are 3.2 to 4.8 kg VS/day/m^3 for mesophilic digesters, and 16 kg VS/day/m^3 for thermophilic digesters.

Typical volumes of mesophilic anaerobic digesters for animal manures in livestock and poultry operations of varying capacities are shown in Table 3.8. These design volumes are based on the foregoing assumed values of design parameters.

Digester Start-up and Operation

Starting with the digester filled with water, only 20 percent of the normal daily manure loading rate should be applied during the first week of digester operation (Fischer, 1979). Thereafter, the daily manure loading rate should be raised by 20 percent each week until the full loading rate is reached. It usually takes 60 to 90 days for a mesophilic anaerobic digester to reach equilibrium conditions. The indications of a stable digester follow (Fischer, 1979; Sweeten et al., 1981).

1. *Methane yield* should exceed 0.37 m^3 of methane per kg of volatile solids destroyed in the digester. The theoretical maximum is 0.50 m^3 CH_4/kg VS destroyed.

2. *Methane concentration* should reach 55 to 65 percent on a volumetric basis.

3. The *destruction of volatile solids* should level off at about 50 percent organic matter destruction (40 to 60 percent range, depending upon retention time and type of waste), as determined by inflow and effluent sampling.

4. The proper *pH range* is 7 to 7.8.

5. *Alkalinity*, an indicator of buffering capacity, should always exceed 1,000 mg/L. Many digesters operate above 5,000 mg/L.

6. Digester slurry should plateau at a *volatile acids concentration* of about 500 to 1,000 mg/L for mesophilic systems and 1,000 to 2,000 mg/L for thermophilic systems. Propionic acid should constitute only a small percentage of the volatile acids.

After equilibrium conditions are reached, it is important to maintain the correct slurry temperature within $\pm$ 2°C of accuracy. A sudden change in animal ration is enough to cause a temporary decrease in gas production (Fischer, Iannoti, and Sievers, 1979). Some disinfectants and feed additives such as antibiotics can cause digester upset. For example, Fischer, Sievers, and Fulhage (1975) experienced decreased

TABLE 3.8 Anaerobic Digester Capacity (Mesophilic) and Expected Gas Yield

No. Animals	Manure Production (kg/day) Total Solids	Volatile Solids	Daily Slurry Input @ 10% TS (m^3)	Total Digester Volume (m^3)	Gas Yield, (m^3/day) Biogas	Methane
Caged Laying Hens (1.8 kg/bird)						
10,000	240	170	2.4	43.0	51	30
Swine (68 kg/head)						
1,000	410	330	4.0	71.0	100	60
Beef Cattle (385 kg/head): Open, Dirt-Surfaced Feedlot						
1,000	3,100	2,040	31.0	560.0	610	370
Beef Cattle (385 kg/head): Confinement Buildings						
1,000	2,700	2,300	27.0	480.0	700	420
Dairy Cattle (635 kg/head)						
500	3,300	2,700	33.0	590.0	800	480

Assumptions: (1) 100% collection of manure; 17-day hydraulic retention time with 5% freeboard; (2) 50% VS destruction; 0.75 m^3 biogas/kg VS destroyed; (3) 20% energy loss for heating and mixing; 60% CH_4 in biogas.

performance caused by lincomycin in a swine ration. The best indicators of digester upset are decrease in biogas production, pH drop, and buildup of propionic acid (Fischer, Iannoti, and Sievers, 1979).

It may be necessary to shut down a digester periodically to remove the settled manure solids and sand that interfere with mixing, and reduce the retention time and effective solids load. When the digester is to be shut down for solids removal, the slurry can be pumped into a covered holding tank, then reloaded after cleanout.

Methane will be produced as long as anaerobic conditions are maintained and pH does not fall below about 6.0. Low pH must be neutralized by adding chemicals such as lime or sodium bicarbonate to raise pH to about 7.0. It may take several weeks to produce methane during start-up or following digester upset. Good digester stability is achieved if input conditions remain fairly constant.

Gas Yields

Specific methane productivity (γ_s) is a measure of both waste degradability and digester efficiency. On the average, about 0.44 to 0.50 m^3 of methane (or 0.75 to 0.80 m^3 of raw biogas) is produced for every kilogram of volatile solids actually destroyed. With volatile solids conversion rates ranging from about 35 to 60 percent for most agricultural residues and wastes, methane yields usually range from 0.16 to 0.31 m^3 CH_4/kg of volatile solids added to the digester.

The volumetric methane productivity (γ_v) is the amount of gas produced per day, per unit of digester volume. For mesophilic digesters it ranges from about 0.5 to 2.0 m^3 of methane/day/m^3 of digester volume, and 3.0 to 3.8 m^3/day/m^3 of digester volume for thermophilic digesters. The volumetric methane productivity rate is an excellent indicator of digester performance.

About 15 to 25 percent of the methane will be needed to provide digester heat (year-round average) and energy for mixing the digester contents. The relative magnitude of heat loss depends on digester configuration, insulation, and heat recovery from digested slurry. Expected methane yields from livestock feeding operations of various sizes are listed in Table 3.8.

Gas-Scrubbing Systems

The degree of biogas cleanup required depends upon the ultimate use of the gas; alternative utilization strategies (Hashimoto et al., 1979a) are discussed later. Water vapor is removed by frostproof condensers and condensate traps to prevent condensation in gas lines and excessive corrosion. Hydrogen sulfide, which also presents a corrosion problem, can be removed by means of an iron sponge (that is, a column of iron-impregnated wood chips) so that biogas can be used in internal combustion engines (Fulton, 1981). The iron sponge can be either regenerated or replaced, since this material is relatively inexpensive. Additional H_2S removal is necessary if the gas is to be sold to natural gas pipeline companies.

Carbon dioxide can be removed by one of several processes, including water scrubbing, membrane separation, phosphate buffer, and regenerative amine absorption. Ashare et al. (1978) evaluated various processes and found that water scrubbing is the most economical method. It consists of bubbling the biogas through water at approximately 3,400-kPa gage pressure to dissolve the carbon dioxide. The heated (regenerative) amine absorption process (Meckert, 1978) and the multiple-stage selective membrane process (Walmet, 1979) have also proved successful in providing pipeline-quality methane gas (37 MJ/m^3) from biogas.

The efficiency of an internal combustion engine burning biogas fuel increases sharply from about 15 to 22 percent as the methane concentration is raised from 60 to 100 percent by carbon dioxide removal. Biogas scrubbing equipment accounts for 30 percent of the total cost of the methane production system for a 10,000-head cattle feedlot if CO_2, H_2S, and water are all removed. If only H_2S and water vapor are removed, the capital cost is only 10 to 12 percent of the total anaerobic digestion system cost (Hashimoto et al., 1979a).

Gas Utilization Strategies

The economical success of methane production is dependent upon efficient on-site utilization or sale of gas and digested slurry. Methane produced from feedlots can provide space heat, steam or dry heat for feed processing, or electricity for feed-processing equipment and lighting requirements. For instance, feed processing at a 30,000-head feedlot would require about 2,600 m^3 of methane per day for heat and 2,400 m^3 per day for electricity, assuming 20 percent efficiency in converting methane to electricity with an internal combustion engine and a generator. This leaves a surplus of about 4,000 m^3 of methane, or 8,300 kWh of electricity, to sell each day. Waste heat captured from the water jacket of the internal combustion engine as well as exhaust gases can be utilized to preheat incoming feedstock slurry and to heat digester contents, thus increasing the methane yield (Hashimoto et al., 1979a; Fischer, Iannoti, and Sievers, 1979).

Five strategies for biogas utilization and marketing at a cattle feedlot were evaluated by Hashimoto et al. (1979a). The most economical strategies for on-site use were as follows: (1) electricity generation after gas scrubbing (CO_2, H_2S, and moisture removal), gas compression to 860 kPa, and storage (one-day capacity); (2) electricity generation after removal of only H_2S and moisture, without compression or storage (CO_2 not removed); and (3) boiler fuel after H_2S and moisture removal, compression to 860 kPa, and one-day storage.

Fulton (1980, 1981) described systems for compressing, storing, and utilizing biogas following hydrogen sulfide removal with a packed-bed iron sponge. The biogas was compressed to 860 kPa and stored in propane tanks with a capacity of 1,000 m^3 at 870 kPa. From the storage tanks, methane gas was used by water heaters to supply heat to the anaerobic digester, a gas air conditioner, and an internal combustion engine that operated an electric generator.

Methane can be compressed and utilized as a vehicle fuel (Henrich, 1979). Feed trucks modified to operate on bottled methane can be driven for about two hours per refill sized at 11 to 15 L of gasoline equivalent. Four feed trucks would use less than 280 m^3 of methane per day.

Feed Value of Centrifuged Slurry Cake

One of the most attractive features of anaerobic digestion is its ability to extract high-protein feedstuffs from the digested slurry. The crude protein content of digested beef cattle manure is about 25 percent (Maciel, 1979; Schellenbach, 1979; Hamilton Standard, 1978). Meckert (1978) obtained a 50 percent removal of digested solids by centrifugation without polymers, while later work has resulted in more than 90 percent removal of suspended solids by addition of polymers followed by centrifugation (James, 1982). Hashimoto, Chen, and Prior (1978) initially recaptured 20 percent of the protein and 35 percent of the solids in digested slurry using a centrifuge, but by sieving the slurry with a #60 mesh sieve followed by centrifugation at 10,000 g for 20 minutes, they were able to capture 63 percent of the solids and 41 percent of the protein. The crude protein content of the recovered solids was 12.5 percent for the sieved solids and 28 percent for the centrifuge cake, indicating that the high-protein biomass is in the finer solids. Centrifuge cake contains about 80 percent moisture, and attempts to feed this product wet have resulted in cattle palatability problems (Maciel, 1979).

Feedlot cattle-feeding trials in California established that the air-dried centrifuge cake had an energy value similar to that of alfalfa (Prokop, 1979). Thermophilic digester cake was nutritionally superior to mesophilic cake. The value of centrifuge cake from digested slurry is estimated at more than twice the methane value (Hashimoto, Chen, and Prior, 1979b). Some researchers estimated the feedstuffs value ranging from $28 to $100 per dry metric ton (Hashimoto, Chen, and Prior, 1979b; Prokop, 1979). However, centrifugation adds 10 to 40 percent to the capital cost of an anaerobic digestion system and is not cost effective for small systems (Hashimoto, Chen, and Prior, 1979b). Nevertheless, Varani (1979) and Maciel (1979) considered the sale of feedstuffs derived from anaerobically digested beef cattle manure to be essential for the economic viability of methane production.

Fertilizer Utilization

Essentially all the plant nutrients originally present in the feedstock remain in the digester slurry. Thus, the amount of land needed for manure disposal could conceivably be about the same as with raw manure, with the added complication of liquid pumping and distribution. However, nutrient recovery by centrifugation, nitrogen losses in the holding pond by ammonia volatilization, and the recycling of liquids for slurry mixing can reduce the land area requirements by 50 to 80 percent. The extent of nitrogen loss from digested slurry in a holding pond was reported by Egg (1979) to be 50 percent during 72 days of storage. Other researchers have found ammonia losses of 5 to 65 percent from land application of anaerobic livestock manure.

Malish, Raley, and Berdoll (1982) measured coastal Bermuda grass yields following land application of anaerobic digester effluent at a swine farm in central Texas. Four adjacent 0.8-hectare test plots received the fertilizer treatments shown in Table 3.9. The anaerobically digested swine manure was pumped from a storage lagoon. It contained 6,000 mg/L total nitrogen (95 percent as NH_4^+), and less than 300 mg/L total phosphorus.

Results from two hay cuttings the first year revealed that both plots treated with digester effluent produced 70 to 73 percent more hay than the control plot, and 13 to 15 percent more than the plot receiving commercial fertilizer treatment. Crude protein analyses of hay were essentially the same among the three treated plots but much lower for the control.

Summary and Conclusions

Anaerobic digestion is a process whereby organic matter is degraded by bacteria to water and gases without atmospheric oxygen

TABLE 3.9 Fertilizer Experiments on Coastal Bermuda Grass

Plot No.	Fertilizer Treatment	Application Rate (kg/ha)		Hay Yield (kg/ha)	Crude Protein (% d.b.)
		N	P_2O_5		
2	Control—no fertilizer	0	0	1,810	7.91
3	Commercial fertilizer	90	45	2,720	11.11
4	Digester effluent[(a)]	74	9	3,080	10.06
1	Digester effluent[(a)]	146	17	3,130	9.78

(a) Anaerobically digested swine manure.
Source: Malish, Raley, and Berdoll (1982).

being present in the reactor. Anaerobic digestion of agricultural wastes and residues produces methane that can be used for energy and a digester effluent that can be utilized as fertilizer or a feedstuff; the process partially stabilizes the organic matter; and it reduces the potential for water and air pollution. A methane production system encompasses feedstock preparation and processing, anaerobic digestion, gas collection, scrubbing and utilization, and digested slurry handling and nutrient recovery.

Anaerobic digestion produces biogas that normally contains 50 to 60 percent methane (CH_4), 40 to 50 percent carbon dioxide, and some trace gases. Raw biogas has an energy content of 19 to 22 MJ/m^3, but can be scrubbed to yield pipeline-quality gas with 37 MJ/m^3. Initial volatile solids concentrations normally are near 8 percent. Volatile solids conversions of 40 to 60 percent are ordinarily obtained, yielding about 0.45 m^3 methane per kg of volatile solids converted.

Biodegradable fractions of most agricultural wastes and residues range from about 50 to 70 percent of the volatile solids, depending upon the carbohydrate composition. Methane is formed principally from acetic acid, which is produced primarily from glucose, following its production from starch and cellulose. The anaerobic digestion process is sensitive to temperature, pH, organic loading rate, solids concentration, and toxic substances (including oxygen, drugs, and metals). Most anaerobic digesters are operated at mesophilic temperatures (28–42°C), a pH of 7.2 to 7.5, influent–total solids concentration of 6 to 12 percent, and a volatile solids loading rate of 1.6 to 3.2 kg/day/m^3.

Numerous mathematical models have been developed to describe the kinetics of anaerobic digestion. These serve to relate the rate of volatile solids conversion to the volumetric methane productivity, which approaches a theoretical maximum of 0.5 L CH_4 per g of VS destroyed. These models are useful for predicting digester performance for a given set of digester and feedstock conditions and for optimizing the system.

Anaerobic digesters can be categorized according to loading frequency (batch, daily, or continuous); temperature (temperate, mesophilic, or thermophilic); degree of mixing (unmixed or plug flow, intermittently mixed, or continuously mixed). Numerous digester configurations have been devised in attempts to optimize methane yield from various feedstocks.

The volumetric rate of methane production has ranged from 0.5 to 2.0 m^3 CH_4/day/m^3 of digester, while for thermophilic digesters it usually exceeds 3 m^3 CH_4/day/m^3. Up to 25 percent of the methane produced is needed to heat the digester contents. Hydrogen sulfide should be removed from raw biogas, using an iron sponge, to prevent corrosion of metal containers and engines. Carbon dioxide removal can be achieved by several processes, the simplest of which is water scrubbing, to yield pipeline-quality gas.

Biogas or methane can be used effectively for space heating, process heat, and generation of electricity in specially equipped stationary internal combustion engines. The digested slurry contains all the plant nutrients that were present in the feedstock and can be an excellent liquid fertilizer. Solids in the digested slurry can be recovered by centrifugation and have some potential as cattle feed.

Among the factors that are important for successful methane production by anaerobic digestion are the following:

(1) collection of high-quality feedstock;
(2) efficient processing of feedstock, including ash removal;
(3) low-cost construction of digester;
(4) efficient recovery and drying of high-protein solids from digested slurry;
(5) heat recovery from internal combustion engines used to convert methane into electricity;
(6) large manure tonnage to achieve economics of scale; and
(7) efficient marketing of all by-products—methane, feedstuffs, fertilizer, and perhaps waste heat and carbon dioxide.

References

Ashare, E., D. C. Augenstein, J. C. Young, R. J. Hossan, and G. L. Duret (1978). *Evaluation of systems for purification of fuel gas from anaerobic digestion*. Engineering Report C00-299-44. Cambridge, Mass.: Dynatech R/D Co.

Ashare, E., and E. H. Wilson (1979). *Analysis of digester design concepts*. U.S. Department of Energy, Document No. C00-2991-42. Cambridge, Mass.: Dynatech R/D Co.

Barker, H. A. (1965). *Bacterial fermentation*. New York: Wiley.

Bryant, M. P. (1979). Microbial methane production: Theoretical aspects. *Journal of Animal Science* 48:93.

Chandler, J. A., W. J. Jewell, J. M. Gossett, P. J. Van Soest, and J. B. Robertson (1980). Predicting methane fermentation biodegradability. *Biotech. Bioeng. Symp.* 10:93–100.

Chen, Y. R., and A. G. Hashimoto (1979). Kinetics of methane fermentation. In C. D. Scott, ed., *Proceedings, Biotechnology in Energy Production and Conservation*. New York: Wiley.

Contois, D. E. (1959). Kinetics of bacterial growth: Relationships between population density and specific growth rate of continuous cultures. *Journal of General Microbiology* 21:40.

Egg, R. P. (1979). Batch load digestion of dairy manure. Paper presented at Texas Agricultural Extension Service, Texas A&M University, and Tarleton State University Seminar on Methane Production from Livestock and Poultry Wastes, Stephenville, Tex.

Fischer, J. R. (1979), USDA/Agricultural Research Service, Columbia, Mo. Personal communication, September.

Fischer, J. R., D. M. Sievers, and C. D. Fulhage (1975). Anaerobic digestion in swine wastes. In *Energy, agriculture and waste management*. Ann Arbor, Mich.: Ann Arbor Science Publishers.

Fischer, J. R., E. L. Iannoti, and D. M. Sievers (1979). *Biological and chemical fluctuations during anaerobic digestion of swine manure*. In press. Columbia, Mo.: USDA/Agricultural Research Service.

Fulton, E. L. (1978), Tarleton State University, Stephenville, Tex. Personal communication, November 16.

——— (1979). Tarleton State University, Stephenville, Tex. Personal communication, August 30.

——— (1980). *Methane generation from animal wastes*. College Station, Tex.: Center for Energy and Mineral Resources, Texas A&M University.

——— (1981). Methane generation: Is it economical? Compendium, Texas Agricultural Extension Service, Texas A&M University System, 26th annual Texas Commercial Egg Clinic, College Station, Tex.

Hamilton Standard (1978). Monfort dirt lot experiments: Status report. U.S. Department of Energy Report No. C00-2952-17.

Hashimoto, A. G., Y. R. Chen, and R. L. Prior (1978). Thermophilic anaerobic fermentation of beef cattle residue. In *Symposium Papers, Energy from Biomass and Wastes*. Washington, D.C.: Institute of Gas Technology.

Hashimoto, A. G., Y. R. Chen, V. H. Varel, and R. L. Prior (1979a). *Anaerobic fermentation of animal manure*. ASAE paper 79-4066. St. Joseph, Mich.: American Society of Agricultural Engineers.

Hashimoto, A. G., Y. R. Chen, and R. L. Prior (1979b). Methane and protein production from animal feedlot wastes. *Journal of Soil and Water Conservation* 34 (1): 16–19.

Hayes, T. (1979). Plug flow reactors for methane production. Paper presented at Seminar on Biogas and Alcohol Production, J. G. Press Co., Chicago, Ill.

Henrich, T. (1979), Dual Fuel Systems, Inc., Montebello, Calif. Personal communication, November.

Hiler, E. A. (1982). *Sorghum for methane production*. Research proposal RF-82-767. College Station, Tex.: Texas A&M Research Foundation, Texas A&M University.

Hill, D. T. (1982a). Design of digestion systems for maximum methane production. *Transactions of the ASAE* 25 (1): 226–30, 236.

——— (1982b). Optimum operational design criteria for anaerobic digestion of animal manure. *Transactions of the ASAE* 25 (4): 1029–32.

Hill, D. T., D. T. Young, and R. A. Nordstedt (1981). Continuously expanding anaerobic digestion: A technology for the small animal producer. *Transactions of the ASAE* 24 (3): 731–36.

James, R. (1982), Thermonetics, Inc., Guymon, Okla. Personal communication, April 29.

Jeris, J. S., and P. L. McCarty (1965). The biochemistry of methane fermentation using C_{14} tracers. *Journal of Water Pollution Control Federation* 37:178–92.

Jewell, W. J., R. J. Cummings, S. Dell'Orto, K. J. Fanfoni, S. J. Fast, E. J. Gottung, D. A. Jackson, and R. M. Kabrick (1982). *Dry fermentation of agricultural residues*. SERI/TR-9038-10, final report. Ithaca, N.Y.: Department of Agricultural Engineering, Cornell University.

Jewell, W. J., H. R. Davies, W. W. Gunkel, D. J. Lathwell, J. H. Martin, Jr., T. R. McCarty, G. R. Morris, D. R. Price, and D. W. Williams (1976). *Bioconversion of agricultural wastes for pollution control and energy conversion: final report*. TID 27164. Ithaca, N.Y.: Cornell University.

Jones, D. D., A. C. Dale, J. C. Nye, and R. B. Harrington (1981). Fiber wall reactor digestion of dairy cattle manure. *Transactions of the ASAE* 24 (5): 1282–86, 1290.

Kemp, M. D. (1978). The potential of biomass conversion for the New Mexico agribusinessmen. M.S. thesis, New Mexico State University, Las Cruces, N.M.

McCarty, P. L. (1964a). Anaerobic waste treatment fundamentals: Chemistry and microbiology. *Public Works* 95 (9): 107–12.

——— (1964b). The methane fermentation process. In H. Henkelkian and N. C. Dondero, eds., *Principles and application in aquatic microbiology*. New York: Wiley.

Maciel, P. (1979). Status of the Imperial Valley biogas project. Paper presented at Seminar on Biogas and Alcohol Production, J. G. Press Co., Chicago, Ill.

Malish, D., D. Raley, and D. Berdoll (1982). Semiannual progress report for the Del Valle hog farm energy integrated farm system. Paper presented at the semiannual review meeting, U.S. Department of Energy, Ithaca, New York.

Meckert, G. W., Jr. (1978). The calorific project. In *Methane production from livestock manure*. Texas Agricultural Extension Service, Great Plains Extension Seminar, Liberal, Kan.

Norstedt, R. A., and M. V. Thomas (1983). *Wood media for fixed bed reactors*. ASAE paper 83-4051. St. Joseph, Mich.: American Society of Agricultural Engineers.

Prokop, M. (1979), University of California, El Centro, Calif. Personal communication, October 29.

Roediger, H. (1967). *Die anaerobe alkalische Schlammfaulung*. 3rd ed. Wasser-Abwasser, 1. Munich and Vienna: R. Oldenbourg.

Sawyer, C. N., and P. L. McCarty (1967). *Chemistry for sanitary engineers*. 2d ed. New York: McGraw-Hill.

Schellenbach, S. (1979). Designing a high quality biogas system. Paper presented at Seminar on Biogas and Alcohol Production, J. G. Press Co., Chicago, Ill.

Stafford, D. A., D. L. Hawkes, and R. Horton (1980). *Methane production from waste organic matter*. Boca Raton, Fla.: CRC Press.

Sweeten, J. M., C. Fulhage, and F. J. Humenik (1981). *Methane gas from swine manure*. Leaflet L-1860. In *Pork Industry Handbook*. West Lafayette, Ind.: Cooperative Extension Service, Purdue University.

Varani, F. T. (1979), Biogas of Colorado, Inc., Arvada, Colo. Personal communication, November 16.

Walmet, G. (1979). Biogas purification with perm-selective membranes. Paper presented at Seminar on Biogas and Alcohol Production, J. G. Press Co., Chicago, Ill.

CHAPTER 4

Biological Conversion and Fuel Utilization: Fermentation for Ethanol Production

CHARLIE G. COBLE, JOHN M. SWEETEN, RICHARD P. EGG, ED J. SOLTES, WILLIAM H. ALDRED, AND DAVID A. GIVENS

Ethanol produced from biomass, a renewable resource, has been widely discussed. Many factions of society have proposed large commitments to the production of ethanol.

Stimulated partially by tax incentives, the ethanol industry has grown at a phenomenal rate since the late 1970s. Fermentation ethanol production capacity in the U.S. rose from about 190 million liters per year in the late 1970s to over 1.7 billion liters per year in 1982, with an additional 380 million liters per year capacity expected to come on line by the end of 1984 (Dyer, 1983).

Several large plants have been built to produce alcohol from grain. Generally, an alcohol production facility is part of a corn wet-milling operation, and alcohol is one of several products that can be supplied by the plant. It has been demonstrated that small-scale facilities (less than six million liters per year) cannot effectively compete for cash sales and are not economically viable for fuel ethanol production unless some unusual condition exists or a locally available feedstock is available at low cost.

Brazil has made major commitments to a fuel-alcohol program. While the overall program is not without problems, production capacity has grown from 1.2 billion liters per year in the late 1970s to 4.2 billion liters per year in 1982 (Dyer, 1983). Brazil has a considerable amount of undeveloped or underdeveloped land available for the production of sugar cane, and a large acreage has already been committed to the crop for sugar production. In addition, large numbers of workers are available at low wage rates. Probably most important is the fact that Brazil produces very little petroleum, and the burden placed on the balance of payments by imported oil is very large.

The program has provided jobs for at least 180,000 workers;

140,000 of these are in agriculture, and the balance work in distillation and distribution. An investment of over $4 billion has been made since 1975, and the foreign exchange savings are about $1 billion per year (Borges, 1983).

The expanded use of alcohols in the U.S. has been primarily to raise the octane rating of unleaded gasoline for automobiles and trucks. Environmental considerations have stimulated engine conversion to unleaded gasoline use. Producing unleaded gasoline from crude oil by conventional methods has resulted in a lower yield of gasoline from a barrel of crude than could be obtained when tetraethyl lead was used to increase the octane number. However, it has been found that when gasoline of a lower octane rating than that required for modern engines is produced from crude, the octane number can be increased by the use of various alcohols or alcohol mixtures (Bolt, 1980). Changes in the octane number of various grades of gasoline by the addition of ethanol is shown in Figure 4.1. It appears that the market for alcohols as octane enhancers is rapidly expanding, and motorists may be burning a lot more alcohol fuel than had been previously considered.

Technology for the production of alcohol is widely known, and the industry can be expanded if economic considerations change. A dramatic increase in the price of crude oil could stimulate increased production of ethanol, provided that production of grain in the United States continues to outpace world markets. In addition, break-throughs in deriving alcohol from nongrain crops or cellulose could drastically lower the cost of the feedstocks and consequently the price of the ethanol produced from these sources. Such breakthroughs will require large research efforts, and many production problems will have to be examined. For instance, the stillage resulting from a very large-scale enterprise could have effects detrimental to the environment unless costly controls are in place. Thus, the present cautious approach to the development of alcohol for fuels appears to be more prudent than the large-scale and rapid expansion of alcohol production from biomass that was favored in the 1973 to 1980 period.

Ethanol Production Processes

The basic process of ethanol production by fermentation involves three general steps. The first step is the enzymatic conversion of starch into fermentable sugars by hydrolysis and saccharification (for sugar

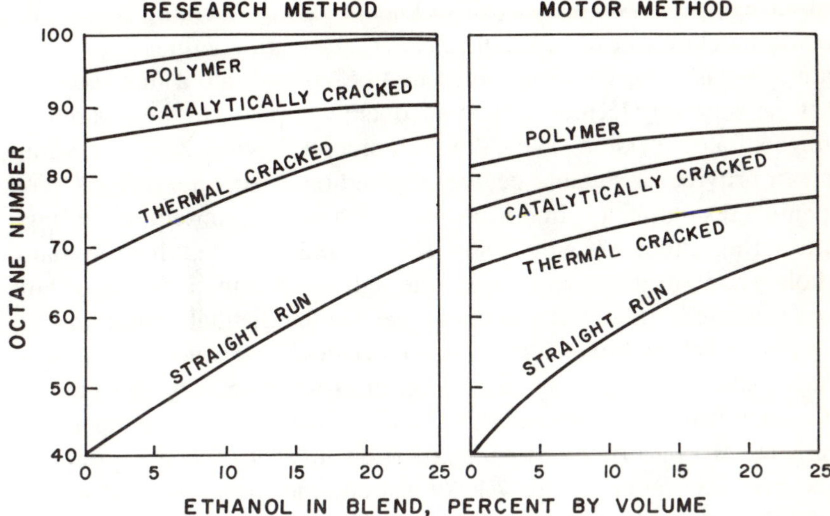

FIG. 4.1. Increase of octane rating in several gasolines with alcohol addition (from Bolt, 1980).

crops this step is generally replaced by sterilization). The second step is fermentation. During fermentation, yeast converts the fermentable sugars in the mash into ethanol and carbon dioxide. This ethanol is separated from the beer by distillation, the final step in the process. The fermented mash, or beer, contains about 8–12 percent alcohol by volume.

Starch Hydrolysis

Before fermentation can take place, the carbohydrates in starchy materials must be broken down into fermentable sugars. Liquefying enzymes, or α-amylases, are used to break down the starch into oligosaccharides such as maltose.

The chemical equation describing the overall reaction is given below, showing maltose as the end product:

$$2nC_6H_{10}O_5 + nH_2O \longrightarrow nC_{12}H_{22}O_{11}$$

$$\text{starch} + \text{water} \xrightarrow{\alpha\text{-amylase}} \text{maltose}$$

$$(1\text{ kg}) + (0.056\text{ kg}) \longrightarrow (1.056\text{ kg})$$

This hydrolysis process, also known as liquefaction, is straightforward and may be carried out as either a batch or continuous process. In either case, the feedstock must first be ground into a fine meal, so that the starch molecules will be adequately exposed to the α-amylase enzyme for conversion. Typically, a hammer mill with a 3.2- to-4.8-mm screen provides a suitable degree of grinding. Using a smaller screen opening results in too fine a grind, which makes subsequent stillage dewatering difficult (Gird, 1980). Hiler (1982) reported that too many whole grain sorghum seeds went through a 4.8-mm screen to permit the fermented grain slurry to be pumped to the distillation column.

For batch hydrolysis, the ground feedstock and water are mixed to form a slurry in a cooking tank. After an α-amylase enzyme is added, the slurry is heated to about 90.0°C and maintained at that temperature for about an hour. Batch cooking is carried out at atmospheric pressure.

For continuous hydrolysis of starch, the grain, water, and α-amylase are fed into a tank and mixed. The slurry is pumped from the tank into a tube reactor, where it is heated to 121–149°C by direct steam injection and pressurized to 550–620 kPa. The residence time in the reactor is about five minutes; the flow velocity is about 0.3 meter per second. A valve at the end of the reactor maintains the required pressure. Upon leaving the reactor, the slurry is cooled in a flash cooler. A continuous hydrolysis system is shown in Figure 4.2.

Several variations of this process have been reported. The process used by Butler and Crombie (1980) involves mixing the grain, water, unnamed chemicals, and dilute acid in a stirred-tank reactor at 188°C and 1,172 kPa with a residence time of 20 seconds. The slurry is then diluted with water and transferred continuously to the single-stage fermentation tank.

In Miles Laboratories' pilot plant (Yang, Grow, and Goldstein, 1981), a primary liquefaction step is carried out as a batch process for one hour at 85°C. The corn mash is then pumped through a steam jet cooker at 140–149°C with a residence time of five minutes. The slurry is flashed to atmospheric pressure and cooled to 85°C. A secondary step is carried out in a series of three stirred tanks, each with a residence time of 20 minutes. Additional α-amylase is added to the first of these tanks.

Downs, Clary, and O'Neal (1982) have done considerable research on starch conversion using a single-screw extruder. Tests run with corn and grain sorghum indicated that the conversion of starch to

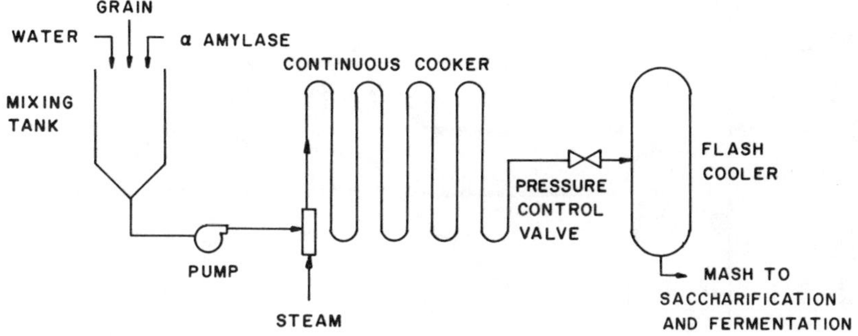

FIG. 4.2. Continuous hydrolysis of mash.

glucose increased with increasing extrusion cooking temperatures. Extrusion cooking at higher temperatures consistently resulted in better conversion than did batch cooking. The effect of temperature on the extrusion of corn is shown in Figure 4.3. Moisture content also influenced starch conversion (Figure 4.4) with maximum conversion occurring at moisture contents of 17 to 19 percent.

Each enzyme manufacturer has its own recipe of optimum conditions for each enzyme, but the temperatures, pressures, and residence times are generally in the ranges described for the batch and continuous conversion processes. Normally a pH of 6.0 to 7.0 is required for efficient conversion.

In the brewing industry, malt is often used as a source of enzymes for breaking down the starch remaining in the malt and in the grain. Malt is obtained by allowing grain, usually barley, to germinate. During germination, enzymes are released to break down the cell walls, and additional amylase is produced. After germination the malt is kilned to fix its properties and is then used as an amylase (Pollock, 1962).

Saccharification

Oligosaccharides must also be broken down into fermentable sugars before fermentation can take place. Invertases, or glucoamylases (saccharifying enzymes), are used to split the oligosaccharides into monosaccharides such as glucose or fructose. These simple sugars can then be converted into ethanol by various yeasts.

The saccharification process does not usually require an actual cooking step; the enzyme is simply added to the slurry. Because the

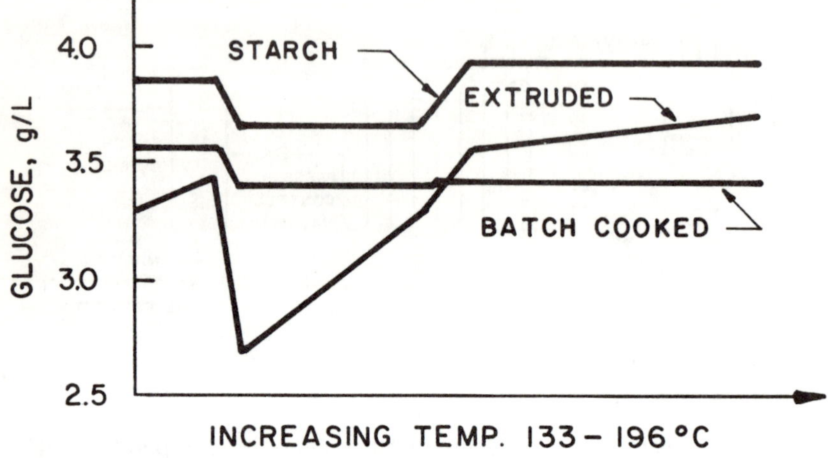

FIG. 4.3. Effect of temperature on extrusion of corn (from Downs, Clary, and O'Neal, 1982).

sugars in such feedstocks as sugar cane and sweet sorghum are readily available for fermentation, the cooking step required for starch is replaced by a pasteurization step to retard the growth of bacteria.

The chemical equations describing the overall reactions for maltose (present in liquefied starch feedstocks) and for sucrose (present in sugar feedstocks such as sugar cane and sweet sorghum) are given below:

$$C_{12}H_{22}O_{11} + H_2O \longrightarrow 2C_6H_{12}O_6$$

$$\text{maltose} + \text{water} \xrightarrow{\text{glucoamylase}} \text{glucose}$$

$$(1\ \text{kg}) + (0.053\ \text{kg}) \longrightarrow (1.053\ \text{kg})$$

$$C_{12}H_{22}O_{11} + H_2O \longrightarrow C_6H_{12}O_6 + C_6H_{12}O_6$$

$$\text{sucrose} + \text{water} \xrightarrow{\text{invertase}} \text{glucose} + \text{fructose}$$

$$(1\ \text{kg}) + (0.053\ \text{kg}) \longrightarrow (1.053\ \text{kg})$$

After cooking is complete, the mash is diluted with water or stillage and cooled to approximately 49°C for saccharification. The mash

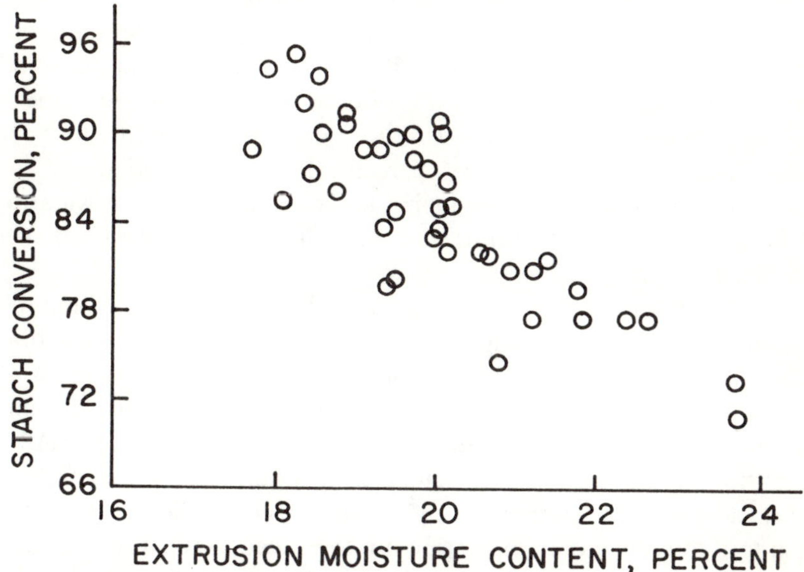

FIG. 4.4. Starch conversion as affected by extrusion moisture content (from Downs, Clary, and O'Neal, 1982).

should be cooled as quickly as possible to retard the growth of bacteria (Mathewson, 1980; Shelton and Rider, 1979a). The glucoamylase is added at a rate of 0.03 to 0.15 percent of initial grain weight. The actual amount varies somewhat with the different enzyme manufacturers. The pH of the mash is normally reduced to 4.0–5.5 with hydrochloric or sulfuric acid for saccharification and fermentation. Some types of enzyme may require a holding period at a specified temperature (Yang, Grow, and Goldstein, 1981).

Enzyme Usage

The optimum conditions and requirements for each brand of enzyme can vary significantly and can be modified somewhat to fit specific conditions. For instance, liquefying and saccharifying enzymes from two different manufacturers were tested by Hiler (1982). The initial recipe for the Canalpha liquifying enzyme and the Gasolase saccharifying enzyme manufactured by Biocon, Inc. (Lexington, Kentucky), involved adding 0.05 grams of Canalpha liquefying enzyme per kilogram of grain when the grain and cooking water were mixed, using 3.7 liters cooking water and 0.7 liters cooling water per kilogram of grain; holding the mash at three different temperatures for specified time periods; and adding 0.45 grams of Gasolase per kilogram of grain at a fermentation temperature of 37°C. After fermentation, the beer alcohol concentration was 5 to 6 percent. This concentration was increased to 6 to 7 percent by reducing the amount of cooking water to 2.5 L/kg and changing the time of the initial addition of the Canalpha liquefying enzyme.

The initial recipe for the Taka-Therm liquefying enzyme and the Diazyme L-100 saccharifying enzyme manufactured by Miles Laboratories (Elkhart, Indiana) involved adding the Taka-Therm at a rate of 1.18 grams per kilogram of grain when the grain and water were initially mixed; using 2.5 L cooking water and 0.7 L cooling water per kg grain; holding the mash at a specified temperature during liquefaction and at another temperature during saccharification; then cooling the mash to a fermentation temperature not to exceed 38°C. The resulting alcohol concentration was increased to 9 to 10 percent by reducing the amount of cooking water to 1.8 L per kg. No other changes in the recipe were necessary.

Thus, the Taka-Therm liquefying enzyme had certain advantages: it required less water and fewer holding periods, and could be added when the grain was added to the cooker. The Biocon Gasolase saccharifying enzyme had the advantages of requiring no holding periods and being added when the fermentation temperature was reached. Therefore, time and temperature control were simplified by using this combination of enzymes. The initial recipes called for cooking at 99°C; however, the cooking temperature was reduced to 91°C which reduced energy usage without any reduction in alcohol yield per unit feedstock.

In the normal recipe, the pH of the mash was lowered from about 6.2 to 5.0 before the saccharifying enzyme and yeast were added. The manufacturers of Gasolase, the enzyme used in the normal recipe, indicated that this pH adjustment was not necessary. In several batches, when pH was not adjusted prior to adding the saccharifying enzymes and yeast, no change in alcohol production was detected; however, severe foaming occurred for about 24 hours in the fermenters, resulting in occasional fermenter overflow (Sweeten, Coble, and Egg, 1982a).

Fermentation

After inversion has been accomplished, yeast must be added to the mash to convert the simple sugars into carbon dioxide and ethanol. The overall reaction for glucose is as follows:

$$C_6H_{12}O_6 \longrightarrow 2C_2H_5OH + 2CO_2$$

$$\text{glucose} \xrightarrow{\text{yeast}} \text{ethanol} + \text{carbon dioxide}$$

$$(1\text{ kg}) \longrightarrow (0.511\text{ kg}) + (0.489\text{ kg})$$

Fermentation of fructose has the same stoichiometric reaction as does glucose.

Distillers' yeast is added to the mash at a rate of about 450 grams of dry yeast per 1,000 liters of mash, an inoculation rate of five to ten million yeast cells per milliliter. The yeasts are sensitive to a number of variables that affect alcohol production: nutritional requirements, sugar concentration, temperature, bacterial infection, and pH (Bothast and Detroy, 1981). Fermentation requires 48 to 72 hours at a temperature of 29° to 35°C. Lower temperatures will slow the activity of the yeast and prolong the process; at higher temperatures the yeast will begin to die. The sugar concentration of the mash should be adjusted with water to about 16 to 24 percent to obtain an ethanol concentration in the beer of 8 to 12 percent. Higher concentrations of ethanol will begin to kill the yeast, resulting in incomplete fermentation; lower ethanol concentrations drastically increase the distillation energy requirements, as shown in Figure 4.5. The ideal ethanol concentration, therefore, is as high as the yeast will tolerate. The optimum pH range for average yeast is 4.5 to 5.0.

Theoretical yields for the overall conversion of starch and sugar to ethanol are 0.568 kg EtOH per kg starch, and 0.538 kg EtOH per kg sucrose. In practice, however, these yields are not attained. Some of the glucose produced is used for yeast cell growth and the production of such by-products as fusel oil and glycerol. Conversion of starch to ethanol is typically 85 to 95 percent (Gird, 1980), and sugar-to-ethanol conversions of 90 percent have been reported (Meade and Chen, 1977).

Batch fermentation is traditional, but considerable progress has been made in continuous fermentation. One continuous process (Figure 4.6) employs six- or seven-stage fermenters, with the conditions

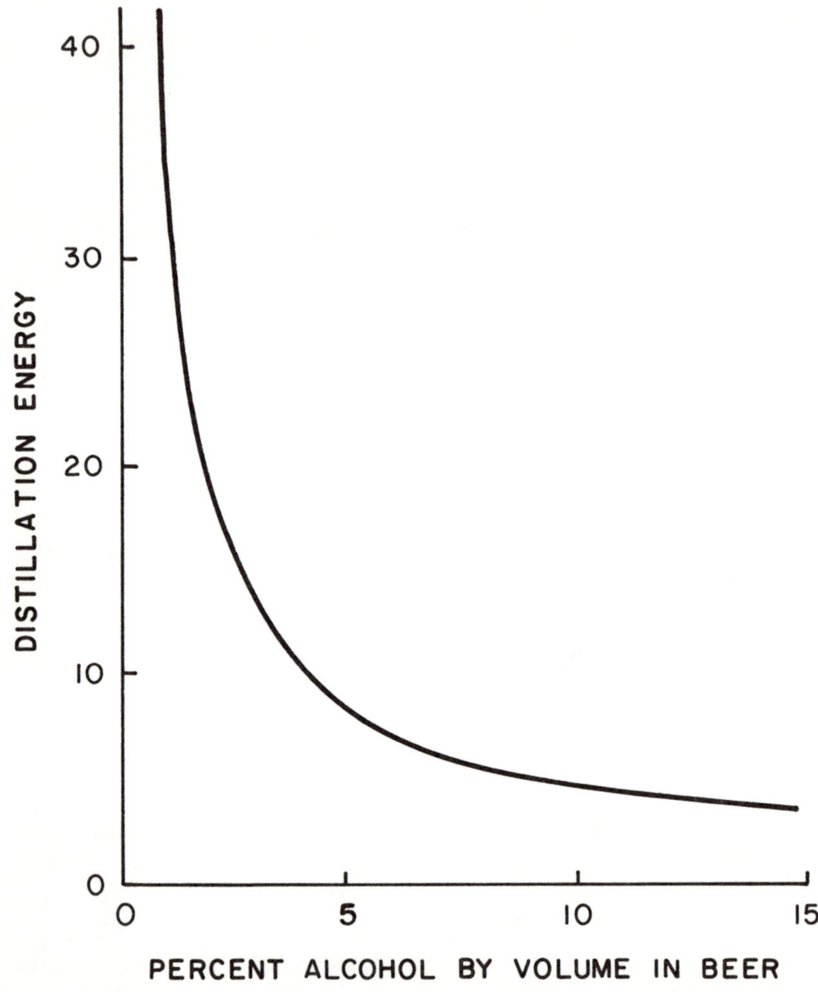

FIG. 4.5. Effect on distillation energy of alcohol concentration in the beer.

for each stage optimized for the corresponding stage in the life cycle of the yeast. The fermenters operate in cascade fashion; steady state is achieved by constant overflow from tank to tank. New or revitalized yeast, fresh mash, and necessary nutrients are added to the first stages. The oxygen and nutrients stimulate growth and reproduction of the yeast. To maximize maturation and the accompanying production of

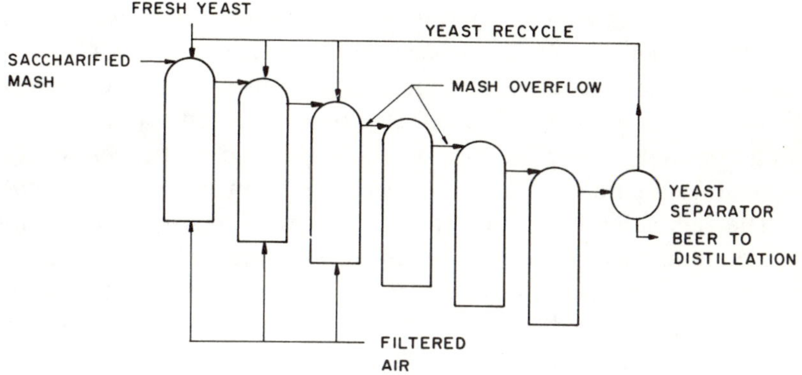

FIG. 4.6. Six-stage continuous fermentation

ethanol and carbon dioxide, oxygen is not provided in the last stages. The final ethanol concentration is about 9 to 10 percent. Antiseptic conditions can be maintained by cooking or pasteurizing the slurry during hydrolysis or saccharification, by filtering the air for the first two or three stages, and by maintaining an ethanol concentration of at least 4 percent (sufficient for antisepsis) in the first stages (Allen, n.d.).

Distillation

Fractional distillation at atmospheric pressure is the most common method for separating ethanol from fermentation beer. Distillation is the physical separation of a mixture into two or more fractions that have different boiling points (Holland, 1975). When this liquid mixture is heated to a given temperature, the resulting vapor will have a higher concentration of the more volatile substance—in this case ethanol, with a boiling point of 78.3°C at 1 atmosphere pressure—than the liquid. The liquid in equilibrium with the vapor will be richer in the less volatile substance. When a hot vapor condenses, the reverse process occurs. The condensing liquid will contain more of the less volatile substance than the vapor from which it condensed. At every given temperature between the boiling points of the pure substances, there is an equilibrium stage beween the liquid and vapor.

The effect of temperature on the equilibrium concentration of ethanol in beer is illustrated in Figure 4.7. If beer with an ethanol content of 10 percent is heated until it begins to vaporize (point x_1), the vapor will have an ethanol concentration of about 53 percent at point y_1. If this vapor is then condensed and vaporized, to point y_2, the vapor will contain about 78 percent ethanol. This process is continued until the desired purity is reached. At point A, the ethanol and water form a constant boiling mixture or azeotrope, and no further separation by atmospheric distillation can occur (Shelton and Rider, 1979b).

The separation of ethanol from beer is made possible in a distillation column that contains a series of equilibrium stages within a temperature gradient. At progressively higher temperatures the stages contain more water and less ethanol. As the temperature decreases, the equilibrium stages contain more ethanol and less water.

Typical distillation apparatus is shown in Figure 4.8. The two sections of a distillation column are often constructed as two separate columns. The section below the beer feed is known as the beer column or stripping column. The function of stages in this section is to increase the recovery of the ethanol. Above the feed stage is the rectifying column. The equilibrium stages in this section serve to increase the purity of the ethanol (Hengstebeck, 1961).

The distillation columns contain devices to increase the vapor-liquid contact area. The perforated metal plate shown in Figure 4.9 is the most common in ethanol distillation. Beer or aqueous ethanol from the higher stage flows down the column through a downcomer and accumulates on the plate. Vapor from the lower stage flows up the column through perforations in the plate and heats and vaporizes the liquid on the plate until equilibrium is reached.

As the liquid flows down the beer column, the ethanol concentration becomes progressively less as the temperature increases. The heat required for the distillation process is normally in the form of steam injected into the base of the beer column.

The vapor that flows up through the rectifying column becomes richer in ethanol until it is removed from the top of the column and condensed. The final concentration is about 190 proof, or 95 percent ethanol.

After condensation, a portion of the ethanol produced is returned to the top of the rectifying column as reflux. This reflux serves as the liquid feed to the final equilibrium stage and is used to maintain the temperature of the final stage. The temperature of the final stage determines the concentration of the final product (Figure 4.7).

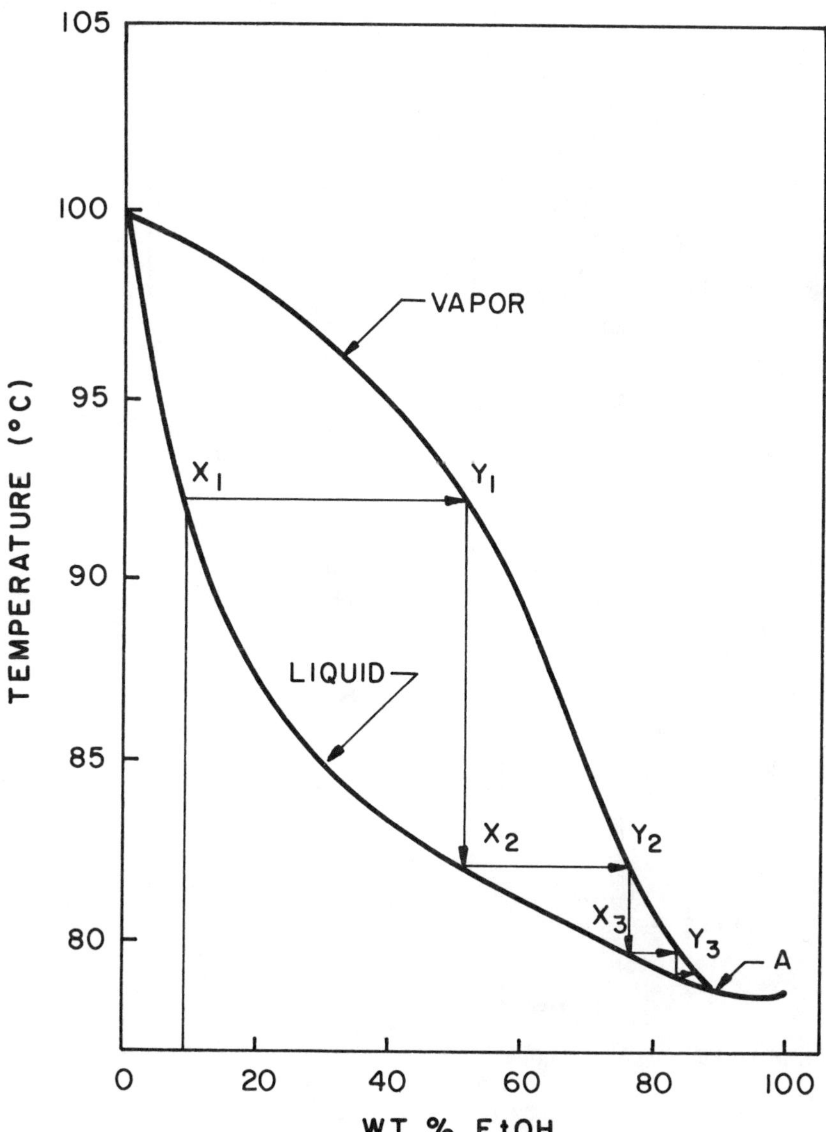

FIG. 4.7. Diagram showing boiling points of ethanol-water mixtures.

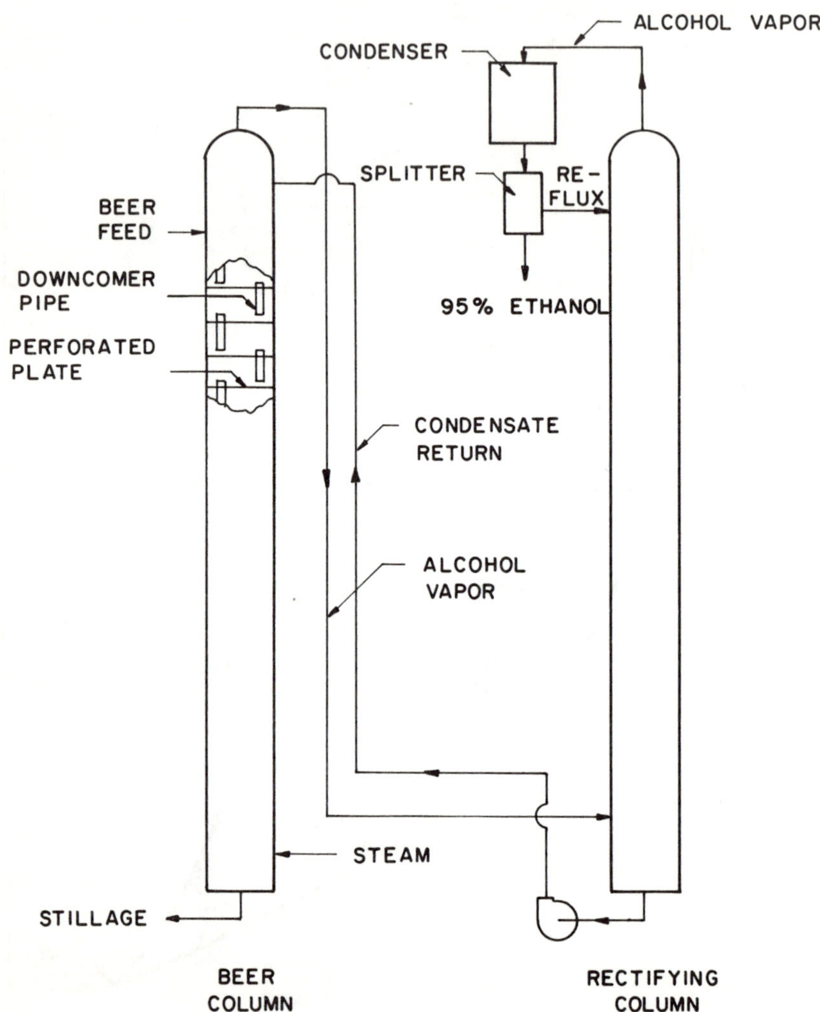

FIG. 4.8. Distillation columns.

Controlling the feed concentration and rate establishes the steam injection rate and reflux rate, equilibrium, and column temperature gradient. In practice, two of the flow rates are set, and the third—usually the beer feed or steam flow rate—is used to control the distillation rate. The reflux rate normally is set in the original column design. Within limits, increasing the reflux rate decreases the number of theoretical equilibrium stages required to achieve the desired separation.

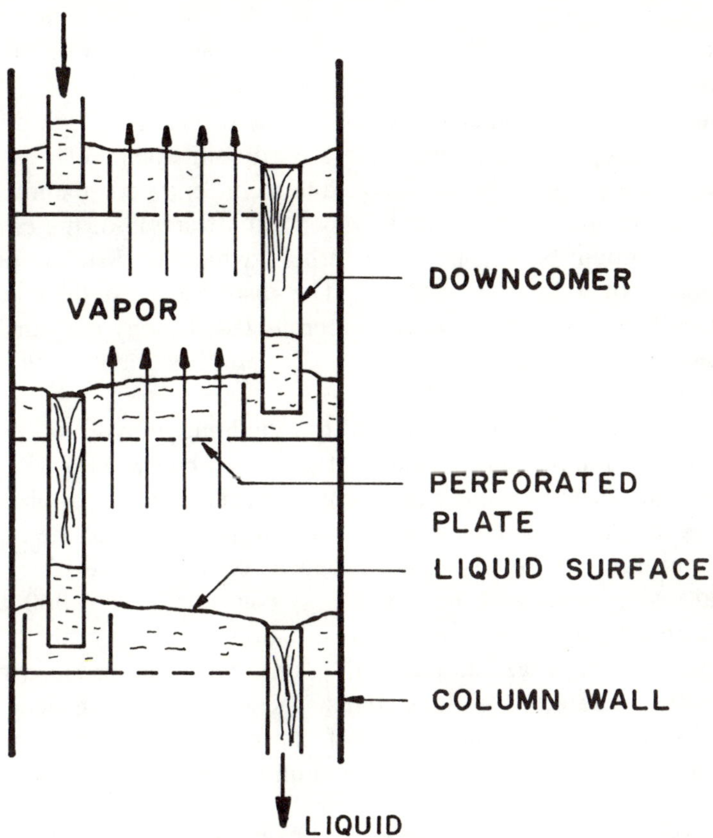

FIG. 4.9. Perforated-plate distillation column.

This in turn reduces the required column height and initial capital cost of distillation. Energy consumption, however, is increased and so, therefore, are operating costs.

In addition to perforated plates, numerous other tray or plate designs are available with varying degrees of efficiency and cost. Packed columns can also be used but generally not with beers containing high solids concentrations because of clogging problems.

Ethanol Dehydration Methods

Several methods have been proposed for removing water from ethanol. Some of these methods are used to obtain anhydrous ethanol either from concentrated ethanol or from dilute ethanol such as fermentation beers; others simply concentrate ethanol to approximately 190 proof.

The conventional method for obtaining an anhydrous product is tertiary azeotropic distillation, using a third component such as benzene or pentane. The benzene is added to a concentrated ethanol solution. The remaining water, the benzene, and a fraction of the ethanol form a minimum-boiling-point azeotrope, which is distilled off to leave pure ethanol in the reboiler. The distillate must then be reprocessed to recover the benzene (Figure 4.10). Energy consumption for concentrating ethanol from 190 to 200 proof is about 9,400 kJ/L (Eakin et al., 1981).

Vacuum distillation is a method for obtaining anhydrous ethanol from a dilute solution in one step. At pressures below 11.7 kPa, the ethanol-water azeotrope does not form, and anhydrous ethanol is obtained as the distillate. However, because of the change of the liquid-vapor equilibrium relations with the reduced pressure, a very large reflux ratio is required, and the total energy consumption (10,300 kJ/L) exceeds that for conventional dual distillation (azeotropic distillation followed by tertiary azeotropic distillation). Since the reduced pressure results in lowered boiling points, vacuum distillation occurs at lower distillation temperatures. This fact would allow cheaper materials such as PVC to be used for column construction rather than stainless steel.

The energy efficiency of distillation to obtain 190-proof ethanol from a dilute solution can be improved by using multi-effect distillation or vapor recompression distillation. Multi-effect distillation employs several columns or effects. The distillate vapor from one effect is used to heat the liquid in the next effect. The pressure must be lower in each successive effect. Vapor recompression distillation uses the distillate vapors to heat the liquid in the reboiler. The vapors must be compressed before being sent to the reboiler; otherwise, no temperature difference between the liquid and vapors would exist, and condensation of the vapors would not occur.

Solvent extraction has been proposed as a means for dehydrating ethanol (Figure 4.11). A carbon dioxide (CO_2) extraction process uses liquid CO_2 at 6,210–6,900 kPa to extract ethanol from 20-proof fermentation beer (Eakin et al., 1981). Ethanol is more soluble in liquid CO_2 than in water. The CO_2-ethanol phase is separated from the water

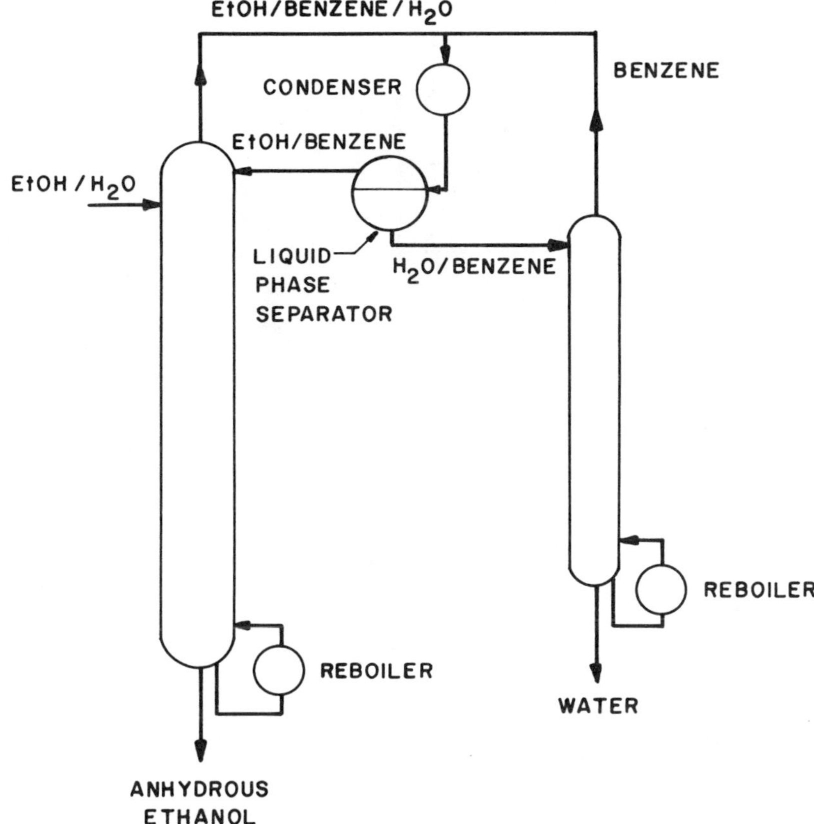

FIG. 4.10. Azeotropic dehydration of ethanol (from Eakin et al., 1981).

phase, then the pressure is reduced to about 4,830 kPa to flash the CO_2, leaving anhydrous ethanol. Similar procedures have been tried with solvents that could be used at moderate pressures.

Low-temperature blending of water-ethanol mixtures with gasoline can be used to produce gasohol in a process very similar to extraction. The water-ethanol mixture is blended with gasoline at −29°C. The water separates from the gasoline to form an ethanol-poor phase. The water phase is then removed and distilled to recover any remaining ethanol. The gasoline phase is used as an ethanol-enhanced fuel. Equi-

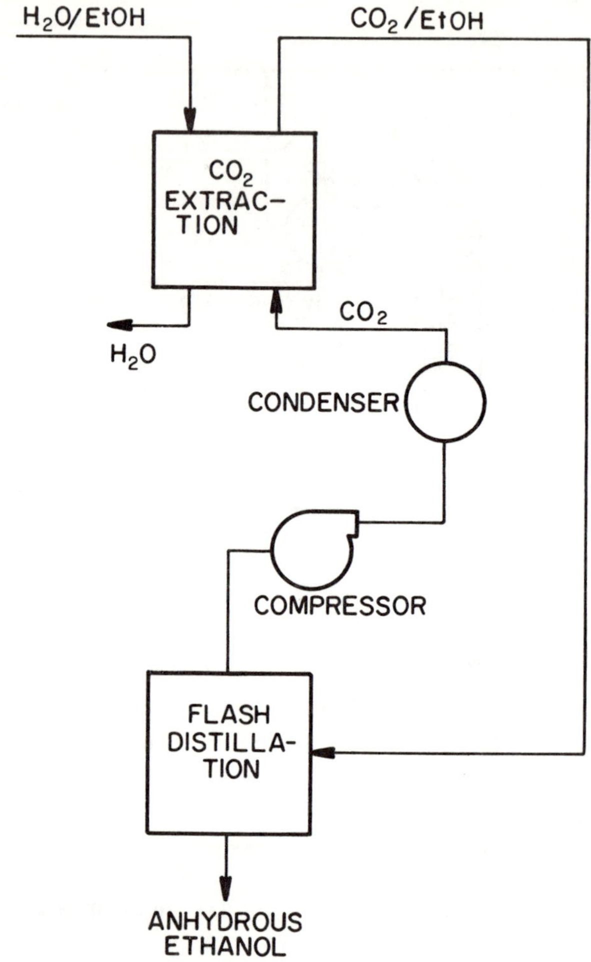

FIG. 4.11. Ethanol dehydration of CO_2 extraction (from Eakin et al., 1981).

librium phase characteristics for this process have not been determined (Eakin et al., 1981).

Water can be separated from ethanol by using molecular sieves. Molecular sieves are crystalline metal aluminosilicates containing a network of interconnecting pores. The pore size in the molecular sieve can be controlled by controlling the crystal formation during manufacturing. Separation is based on the selective adsorption of molecules onto the pore walls of the sieve.

For separation of water from ethanol, a sieve with a 3-Å effective pore diameter is used. Water, with a molecular diameter of less than 3 Å, is adsorbed by the sieve; ethanol, whose molecules are larger than 3 Å, cannot be adsorbed (Gasohol U.S.A. 1981).

The component of a mixture for which the sieve has the greatest affinity is removed from the mixture until the sieve is saturated. Although molecular sieves can be used to remove ethanol from fermentation beers, water is usually removed from high-proof ethanol obtained by conventional distillation. Regeneration is the process by which the column is made ready for reuse; it is accomplished by shifting the column equilibrium so that the adsorbed material may be removed. This shift may be carried out by raising the column temperature or by reducing the column pressure. In either case a purge gas is needed to strip the column of the adsorbed material.

Ethanol dehydration can be done as a batch or continuous process. Continuous dehydration requires two columns, one of which separates the ethanol and water while the other is being regenerated. The ethanol-water mixture may be passed through the column as a liquid or as a vapor. Liquid velocities should not exceed 0.076 m/min and are usually less than 0.03 m/min (Gasohol U.S.A., 1981). The energy required to concentrate 190-proof ethanol to 200 proof is usually between 1,310 and 1,750 kJ/L.

One molecular sieve dehydration system has reported energy requirements of 120–160 kJ/L (Ad-Pro Industries, undated). The sieve is regenerated with vacuum rather than heat. About 20 percent of the anhydrous ethanol vapor produced is passed back through the sieve bed as a purge gas to remove the water. This purge flow is then returned to the reflux of the rectifying column for redistillation. The energy usage reported is for electricity to operate the vacuum equipment, and steam to superheat the inlet vapor stream. Reasons for the low energy requirements are said to be that vacuum regeneration does not require the large amounts of energy normally associated with heat regeneration. The reported figures do not include the energy required for redistilling the purge stream. Several of these units have been installed in commercial operations and appear to be performing satisfactorily.

The use of desiccants such as calcium oxide or fermentable grains has been explored. Ladisch and Dyck (1979) performed laboratory tests on a number of materials to determine their potential for ethanol dehydration. The results are shown in Table 4.1. Both starch and cellulose products were found to be effective dehydrating agents.

The energy requirements of various existing and proposed ethanol

TABLE 4.1 Comparison of Alcohol Dehydrating Agents

Material	Ethanol (%)	
	Starting	Dehydrated
Cornstarch	73.7	99.0
Sucrose	72.5	90.7
Corn	77.0	97.7
Auicel (microcrystalline cellulose)	88.9	98.6
Whatman CF-11 cellulose	88.8	96.4
Buckeye CM cellulose	84.8	99.8
Corn residue	85.2	92.0
NaOH	80.7	97.6
$CaSO_4$	90.1	98.0

Source: Ladisch and Dyck (1979).

dehydrating processes are summarized in Table 4.2. When some of the new technologies become commercially developed, ethanol separation energy utilization can be substantially reduced.

Feedstocks

Any plant material that contains sugar or starch can be used as a feedstock for producing ethanol by fermentation. From a practical standpoint, however, grain is the predominant feedstock—at least in the United States. The Brazilian ethanol industry is based predominantly on sugar cane, but sugar cane is essentially a tropical crop and cannot be widely grown in the U.S.

Grain

Grain is the primary fermentation feedstock in the U.S. for three fundamental reasons. First, it is widely available: cereal grains account for nearly 50 percent of the harvested acreage of all field crops in the U.S. (Lockeretz, 1980). Second, the technology for harvesting, handling, and processing grains for fermentation is well established. Third, grain is a storable commodity. To minimize the unit capital costs of ethanol production, a plant must operate year round; hence, since most crops are seasonal, storage of the feedstock is necessary. Grain can be

TABLE 4.2 Projected Energy Requirements for Ethanol/Water Separation Processes

Type of Separation	Ethanol Concentration		Process	Energy (kJ/L)
	Initial	Final		
Complete	10%	100%	conventional "dual" distillation	7,640
			CO_2 extraction	2,230–2,790
			solvent extraction	1,000
			vacuum distillation	10,310
To azeotrope	10%	95%	conventional distillation	5,020
			vapor recompression	1,780
			multi-effect vacuum	2,100
Azeotropic	95%	100%	conventional azeotropic distillation	2,620
			dehydration	335 [(a)]
			low-temperature blending with gasoline	836 [(b)]
			molecular sieve	1,310–1,750
	90–95%	100%	molecular sieve	125–151 [(c)]
Other	5%	10%	reverse osmosis	139

Source: Eakin (1981).
(a) For drying with CaO, energy requirements using fermentable grains would be considerably less.
(b) Results directly in production of gasohol.
(c) Information letter from Ad-Pro Industries, Inc. (undated).

stored indefinitely with little loss of carbohydrate quality, whereas most other sugar and starch crops—such as sweet sorghum or potatoes—begin to deteriorate during storage, with a resulting reduction in ethanol yield.

A solution to this problem would be to use one such feedstock while it is in season, and operate the plant with grain for the remainder of the year. The problem with this approach, however, is that the processing equipment used for grain may not be compatible with the processing requirements of nongrain feedstocks. Thus, although several research programs are underway to develop fermentation pro-

TABLE 4.3 Alcohol Potential from Leading U.S. Grains

	Corn	Wheat	Sorghum
Harvested area, 1979:[(a)]			
% of all field crops	21.2	18.7	3.9
Total production, 1969–79 average:			
million metric tons	147.3	48.7	19.5
Potential ethanol yield:[(b)]			
L/ha	2,290	917	823
L/metric ton	387	343	373

(a) From Lockeretz (1980).
(b) From Gird (1980).

cesses for nongrain feedstocks, the commercial fermentation ethanol industry still depends predominantly on grain.

The three leading grains produced in the U.S. are corn, wheat, and grain sorghum; these account for nearly 44 percent of the total acreage of field crops harvested in 1979 (Lockeretz, 1980). Production statistics and estimated ethanol yield are shown in Table 4.3. The expected ethanol yields for these grains are 387 liters/tonne for corn, 343 liters/tonne for wheat, and 373 liters/tonne for grain sorghum (Gird, 1980). On a unit land area basis, corn is the leading ethanol producer.

Corn is by far the most common fermentation ethanol feedstock. A small amount of grain sorghum is used in the Southwest, but very little wheat is used for ethanol production. Wheat has historically been priced about 24 percent higher than corn, making it a more expensive feedstock (Lockeretz, 1980). Average prices for these grains for 1969 to 1983 is shown in Table 4.4. If this price trend continues, wheat will see little additional use as an ethanol feedstock. The price of grain sorghum, on the other hand, has been about 8 percent less than that of corn, but the ethanol yield is slightly lower. Sweeten, Coble, and Egg (1982a) reported yields of up to 364 liters/tonne from grain sorghum in a pilot-scale ethanol plant. This yield is about 6 percent less than the 387 liters/tonne common from corn. Even with a lower ethanol yield, grain sorghum has the potential for increased utilization as an ethanol feedstock if it continues to be priced more favorably.

For the near future, however, corn will probably remain the pre-

TABLE 4.4 Average Prices of Leading Grains

	Corn	Wheat		Grain Sorghum	
Year	$/tonne	$/tonne	% difference from corn	$/tonne	$ difference from corn
1969	44	46	4.5	41	− 6.8
1970	48	49	2.1	42	−12.5
1971	50	50	0.0	47	− 6.0
1972	46	58	26.0	45	− 2.2
1973	74	116	56.8	71	− 4.1
1974	115	165	43.4	101	−12.2
1975	106	135	27.4	95	−10.4
1976	98	115	17.3	88	−10.2
1977	80	84	5.0	69	−13.8
1978	83	104	25.3	76	− 8.4
1979	93	129	38.7	86	− 7.5
1980	122	154	26.2	116	− 4.9
1981	98	144	46.9	94	− 4.1
1982	105	140	33.3	99	− 5.7
1983	133	140	5.3	117	−12.0
			Avg. 23.9		Avg. − 8.1

Source: Lockeretz (1980).

dominant ethanol feedstock in the U.S. because of its widespread availability. Utilization of other grains will depend on relative cost per unit of ethanol yield.

Sweet Sorghum

Considerable interest has recently been shown in using sweet sorghum as an ethanol feedstock. The primary reason is that sweet sorghum has a potential ethanol yield of about 4,000 liters per hectare (Bryan et al., 1981a). This compares with grain ethanol yields of 2,290 liters per hectare for corn, 917 liters per hectare for wheat, and 823 liters per hectare for grain sorghum (Table 4.3). Sweet sorghum is a relatively hardy plant and can be grown almost anywhere that has about a three-month growing season. Water and soil nutrient requirements are less for sweet sorghum than for grain crops.

Another advantage of sweet sorghum is that it is a sugar, rather than starch, crop. The starch conversion (cooking) step can thus be eliminated from the fermentation process, resulting in reduced energy consumption and lower costs; however, research has indicated that a sterilization process may be required to replace the starch conversion step, thereby nullifying this advantage. Rein et al. (1982) investigated the effects of cooking sweet sorghum juice for 30 minutes at 60°C and 85°C prior to fermentation. They found that after laboratory fermentation, the cooked samples resulted in conversion efficiencies greater than 70 percent; most uncooked samples had conversion efficiencies of less than 40 percent. The results were attributed to different microbial conditions between the cooked and uncooked samples.

Two major problems have prevented the widespread utilization of sweet sorghum as an ethanol feedstock: the tendency for the sugar to deteriorate during storage, and the difficulty in extracting a high percentage of the sugar from the stalks. Means for overcoming these problems are the objectives of current research.

Eiland, Clayton, and Bryan (1982) conducted a study to determine the effect of harvesting methods on fermentable sugar losses over a one-week storage period. Sweet sorghum harvested and stored as whole stalks and billets approximately 0.6 m long had no significant sugar loss over a one-week period. Chopped sorghum harvested with a forage harvester, however, showed a rapid loss of sugar during the first 24 hours. Freshly expressed juice begins spoiling after 5–12 hours (Dasschel, Maudt, and McCarty, 1981). Effective means for storing sweet sorghum still need to be developed to overcome the deterioration problem.

The second major objective of current research is to develop efficient, cost-effective sugar extraction techniques. Commercial sugar mills are capable of extracting more than 90 percent of the sugar from sugar cane; however, the necessary equipment is capital intensive, and large amounts of land dedicated to a sugar crop are required in order to justify the investment.

Research has focused on alternative methods of low-cost sugar extraction. Reidenbach and Coble (1982) compared three methods of sugar extraction: chopping the whole stalks for fermentation, extracting the juice with a three-roller mill, and extracting the pith by the Tilby process. (In the Tilby process, sorghum or sugar cane billets of approximately 50 cm in length are fed through a machine that splits the billets lengthwise. Serrated rollers then separate the pith from the rind, and the pith is used as a fermentation feedstock.) The chopped sorghum and the Tilby pulp were covered with water and heated for sterilization; yeast was then added, and the material fermented and distilled. No water was added to the juice before sterilizing and fermenting and distilling.

The chopped sorghum gave the highest production efficiency—74 percent; however, the quantity of water necessary to cover the solid material resulted in low alcohol concentrations in the beer, greatly increasing distillation energy requirements. Roller milling resulted in an extraction of 55 percent of the juice and an overall ethanol yield of 44 percent. The Tilby pulp yielded only about 42 percent of the potential alcohol (estimated on the basis of the sugar content of the stalk).

Bryan et al. (1981a) evaluated the juice extraction efficiency of a three-roller mill for eight varieties of sweet sorghum. Juice extraction efficiency ranged from 29.9 to 37.5 percent. Sugar recovery ranged from 37.0 to 48.4 percent (Table 4.5).

A juice harvester, constructed by adding a roller mill to a forage harvester, resulted in a sugar extraction efficiency of only 13 percent (Nuese and Hunt, 1982). The low figure was attributed to mill wear and old age.

The foregoing results indicate that the greater ethanol production potential of sweet sorghum over grain crops has not yet been realized in practice. However, if the technical problems just discussed are solved, sweet sorghum may make a great impact on the ethanol industry.

Sweet Potatoes

Sweet potatoes are another crop with a high potential alcohol yield—nearly 3,000 liters per hectare (Gird, 1980). Egg, Coble, and O'Neal (1982) reported ethanol yields from fermented sweet potatoes ranging from 77 to 137 liters/tonne. When conducting studies in a small-scale ethanol production facility designed for grain, they found that water dilution was required to permit pumping and handling, resulting in a maximum ethanol concentration in the beer of only 6.1 percent by volume.

Badger et al. (1982) reported ethanol yields of 83 to 132 liters/tonne for three varieties of sweet potatoes. Similar dilution problems were encountered.

TABLE 4.5 Field Data, Stalk Analysis, and Sugar Recovery (3-Roll Mill) for Sweet Sorghum

Variety	Plant Population (10^3/ha)	Stalk Weight (t/ha)	Dry Matter (%)	Sugar Content (%)	Theoretical Sugar Yield (%)	Juice Extraction (%)	Sugar Recovery (%)
Brandes	178	30.4	24.5	11.2	3.41	36.5	48.4
Keller	268	42.8	34.2	13.8	5.89	31.4	45.7
MN1500	155	44.2	30.7	13.1	5.80	31.7	40.4
Ramada	102	41.2	33.1	14.0	5.78	32.7	43.1
Rio	175	28.7	31.4	13.6	3.88	29.9	36.4
Roma	100	31.7	31.9	14.0	4.41	31.4	38.6
Theis	111	31.0	27.4	14.1	4.34	37.5	37.0
Wray	214	28.7	32.3	14.3	4.09	36.0	46.6

Source: Bryan et al. (1981a).

Miscellaneous Feedstocks

Numerous other feedstocks are available as ethanol feedstocks. Badger et al. (1982) have conducted fermentation trials for a number of crops. Their results are shown in Table 4.6. An ethanol yield of 63 liters/tonne was reported by Hiler (1982), using cull peaches as a feedstock. Most of these nongrain crops have limited practical application as ethanol feedstocks at the present time because of technical processing problems or limited cultivation.

Lignocellulosic Feedstocks

Although most current activity in the production of ethanol for liquid fuel is concerned with the direct fermentation of starch- or sugar-containing crops (Soltes, 1980), research is expanding into the use of lignocellulosics, such as agricultural and forest wastes. It has been stated often that plant cellulose is the most abundant renewable resource on earth. At approximately 50 percent of all biomass, annual production is about 50 billion tons (Goldstein, 1981a). Because it is potentially available in perpetuity—assuming that plants are managed properly—and because it is abundant, cellulose should be an increasingly important resource for our material, energy, and food needs. For the most part, however, it remains a somewhat elusive resource.

Cellulose, a polymer of glucose, is generally found in nature in a plant cell-wall substance called lignocellulose. Lignocellulose is an intricate matrix of three components: cellulose, lignin, and hemicellulose. Conditions used to liberate cellulose from this matrix are also often responsible for degrading cellulose; therefore, it is difficult to obtain high yields of "pure" cellulose (Soltes, 1983).

As isolated polymers, celluloses can be hydrolyzed by acids or cellulolytic enzymes to yield glucose. However, only acids can readily hydrolyze cellulose as it exists in a native lignocellulosic matrix, and the conversion of cellulose into glucose by mild acid hydrolysis requires a high activation energy because the ordered, so-called crystalline, structures of cellulose present in lignocellulose contribute to slow hydrolysis rates. Moreover, during the time required for acid hydrolysis, the glucose can be degraded, severely limiting final glucose yields (Goldstein, 1983).

Other components in lignocellulose and/or their arrangement in the lignocellulosic matrix appear to have an inhibitory effect on cellulose accessibility for conversion into glucose by the cellulolytic en-

TABLE 4.6 Summary of Selected Feedstock Fermentation Trials, 1981

Feedstock	Variety	Feedstock Amount (kg)	Feedstock Yield (kg/ha)	% Moisture	% Sugar (dry basis)	% Starch (dry basis)	% Ethanol in Beer (vol.)	Fermentation Yield (L/kg dry matter)[a]	Fermentation Yield (L/kg feedstock)[a]	Theoretical Yield (L/kg feedstock)[a]	Conversion Efficiency (%)	Yield (L/hectare)[a]
Corn (typical)		454	9,410[b]	14.9	3.6	66.0	8.0	0.403	0.342	0.387	88.0	3,220
Apples	Yellow Delicious (cull)	919		88.5	2.4	9.0	3.3	0.373	0.043	0.060	72.0	
Apples	Red Delicious (cull)	816		86.5	2.8	10.7	4.7	0.406	0.055	0.060	91.0	
Beets, fodder	Field run mixture[c]	908					1.26		0.013	0.092	14.5	
Beets, sugar	Field run mixture[c]	667		64.1			1.76	0.066	0.024	0.092	26.0	
Jerusalem artichokes		907	27,200	77.8	5.1	15.6	3.4	0.301	0.067	0.125	80.0	1,810
Jerusalem artichokes		23	27,200		5.1	15.0	8.5		0.142	0.125	100.0	3,850
Potato (Irish)	Field run mixture[d]	907	18,010	80.9			71.0	0.384	0.074	0.126	84.0	1,330
Potato (Irish)	Cull & marketable mixture[d]	2,022	18,010	83.5			6.5	0.451	0.075	0.126	87.0	1,350
Potato (sweet)	Triumph white	912	22,800	66.0	1.3	32.5	7.8	0.365	0.124	0.143	86.8	2,830
Potato (sweet)	Carver	816	28,000	71.9			5.1	0.297	0.084	0.143	58.8	2,340
Potato (sweet)	Red 1869	455	21,400	70.3			7.1	0.367	0.109	0.143	76.3	2,320
Potato (sweet)	Carver	45	28,000	71.9			10.5	0.469	0.132	0.143	92.3	3,680
Sweet sorghum juice	Rio and MER 71-7	1,325[e]	6,060[e]		18.3[f]		9.5		0.096[g]		89.0	1,430
Sweet sorghum juice	Rio and MER 71-7	1,488[e]	6,060[e]		18.3[f]		7.8		0.090[g]		65.0	1,200
Water chestnuts		403	20,200	82.0			5.5	0.442	0.079			1,600

Source: Badger et al. (1982). Original data in English units.

(a) Liters 200-proof ethanol.

(b) 13.6 m^3/hectare assumed.

(c) Mono HY T8, Barwi, Mono Blanc #1 and #2, Barsien, USH 20, Mono Rosa, Mono Hybrid, Capax, AAN 23, and Bario varieties.

(d) Atlantic, Superior, Red Pontiac, Kennebec, La Chipper, and Le Soda varieties.

(e) Liters of juice.

(f) Wet basis.

(g) Liters 200-proof ethanol per liter of juice.

zymes of microorganisms. This inhibition is found in cellulolytic enzyme isolates and also parent microorganisms, including the microflora used by ruminants for digestion. Ruminants, despite an ability to convert cellulose into glucose, have problems in accessing the cellulose present in the lignocellulosic tissues of many plants. These plant tissues are composed of additional substances and/or ordered in such a way that much of the cellulose present is not readily available for utilization.

The conversion of cellulose to glucose is usually considered the first step for potential large-scale utilization of cellulose. If acceptable yields of glucose could be obtained through either acid or enzyme hydrolysis processes, glucose could be readily fermented to ethanol in high yield, using commercially proven processes. This ethanol could then be the basis for a chemical industry (Goldstein, 1974, 1977, 1978). Suggested approaches include the dehydration of ethanol to yield ethylene, and conversion to butadiene by means of processes that were once proven commercially but made obsolete by formerly inexpensive petroleum. Our dependence on petroleum for chemicals is largely due to its content of olefins. The production from ethanol of ethylene (the largest-volume organic chemical used as a building block for petrochemicals and plastics) and of butadiene (used in the production of synthetic rubber) are obvious ways to decrease our dependence on petroleum. Lactic acid from glucose can be converted to acrylic acid and acrylates. In addition, glucose can be readily fermented into a wide assortment of antibiotics, chemicals, and vitamins (Seeley, 1976). Glucose fermentation, one of man's oldest processes, is nevertheless in its infancy, relative to its potential.

Acidic and Enzymatic Hydrolysis of Cellulose. It has been known for over 150 years that cellulose can be converted to glucose by acid hydrolysis. The basic dilute and concentrated acid processes that have been developed over the years are well summarized by Goldstein (1981b). There are yield limitations, generally 50 to 60 percent, in high-temperature dilute acid processes because of the slow hydrolysis of highly ordered cellulose with subsequent degradation of product glucose (Goldstein, 1983). The use of concentrated acids requires corrosion-resistant materials and acid recovery cycles, although yields of product glucose can approach or exceed 85 percent. Research continues today toward the development of cost-effective processes for both approaches. Most of the activity has involved sulfuric and hydrochloric acids, although there has been work with others (Hall, Saeman, and Harris, 1956). There is renewed interest in the use of liquid hydrogen fluoride (Selke, Hawley, and Lamport, 1983) because of the potential for nearly complete recovery of glucose.

Enzymatic hydrolysis of cellulose, or at least the idea that enzymes can be used industrially for this purpose, is a much newer activity for cellulose conversion scientists. Most of this activity has focused on the cellulase complex isolated from the fungus *Trichoderma reesei*, although in very recent years, a host of fungi, bacteria, and yeasts or their enzyme isolates have been receiving attention. Research in the late 1970s was spurred by the U.S. Department of Energy's gasohol programs, most of it based on *T. reesei* cellulase to produce glucose for fermentation by *Saccharomyces cerevisiae* to produce alcohol. Like the acid hydrolysis of cellulose, enzymatic hydrolysis of cellulose is complex, and many obstacles still remain to its routine application on an industrial scale. Although isolated celluloses can be readily hydrolyzed, the major problem appears to be some structure in lignocellulosics that inhibits access of the cellulose by cellulase enzymes.

For both acidic and enzymatic hydrolysis, as for ruminant nutrition, the susceptibility of lignocellulose to digesting agents must be enhanced; physical barriers must be removed. The facile solution appears to be chemical pulping to removed hemicelluloses and lignin. However, although it has been demonstrated that cellulose does become much more accessible to acids, enzymes, and rumen microflora with such treatment, the expense of pulping processes is generally thought to be excessive for this purpose.

Additionally, there are physical methods to enhance the accessibility of cellulose in lignocellulosics. Most of these relate to comminution to increase surface areas and improve the porosity of lignocellulosics to digestive agents. Explosive decompression is another approach receiving attention today (Marchessault and Malhotra, 1983; Muzzy et al., 1983). The use of sulfur dioxide, ammonia, steaming, and other physicochemical pretreatments for enhancing cellulose digestion have been well documented, for purposes of both cellulose hydrolysis (Goldstein, 1981a) and ruminant nutrition (Huber et al., 1983; Han, Catalano, and Ceigler, 1983).

Several options are being investigated for the recovery of cellulose itself, rather than its derivatives. One method is the use either of solvents that can selectively dissolve cellulose out of the lignocellulosic matrix (Turbak, 1983) and thus make it available for use in a clean, lignin-free state, or of appropriate swelling and derivatizing agents to remove cellulose as a soluble derivative (Durso, 1983). Alternatively, pulping with organic solvents such as ethanol (Sarkanen, 1980; Klausmeier, 1983; Lipinsky, 1983) exhibits promise for the selective removal of lignin and hemicelluloses from lignocellulosics, leaving reactive cellulose as a product.

Hydrolysis and Fermentation of Hemicelluloses. There are two basic hemicellulose types: glucomannans (polymers of glucose and mannose), found in conifers; and xylans (polymers of xylose), found in most other plants (Thomson, 1983). Because of the more ubiquitous nature of the xylan polymer, and because this hemicellulose plays an important role in animal nutrition, it has been more thoroughly studied for potential utilization in fuels and chemicals.

Xylan contains labile acetate groups that are readily hydrolyzed into acetic acid (the source of the vinegar aroma of felled hardwoods and decaying agricultural products). Because of this acetic acid, xylan is subject to autohydrolysis upon steaming, yielding the sugar xylose and smaller amounts of other sugars. Autohydrolysis has been used as a pretreatment for agricultural products to improve both ruminant digestibility and enzyme accessibility (Soltes, 1983), and xylose is itself used in the production of furfural and of xylitol, a nonnutritive sweetener.

In acid hydrolysis of cellulosic materials, a prehydrolysis step is normally used to remove xylan as xylose, prior to stronger hydrolysis conditions to hydrolyze cellulose to glucose. Most cellulolytic enzyme systems, including the *T. reesei* enzymes, contain hemicellulases (xylanase and mannanase) with the ability to hydrolyze the hemicelluloses to the parent monomer sugars. Recently, it has been found that xylose can be fermented to ethanol, although by an indirect route: the yeast *Pachysolen tannophilus* can isomerize xylose to xylulose, which is fermentable with normal *Saccaromyces cerevisiae* yeast to produce ethanol (Detroy, Cunningham, and Herman, 1982; Jeffries, 1982). The xylose isomerase is identical to that used to isomerize glucose to fructose in the production of high-fructose corn syrup (HFCS). A number of laboratories around the country are attempting to incorporate the xylose isomerase enzyme into *S. cerevisiae* or similar ethanol-producing yeasts for direct conversion of xylose to ethanol.

The general trends for utilization of lignocellulosics are toward complete fermentation of hemicellulose and cellulose to ethanol. There has been increasing attention paid to different fermenting microorganisms (such as *Torulopsis wickerhamii*) with abilities to convert cellodextrins to alcohol (Lastick, Spindler, and Grohmann, 1983). Cellodextrins are oligomers of glucose (very short cellulose chains). It is anticipated that work of this nature will eventually permit the direct conversion of cellulose to ethanol.

Cellulose is a versatile raw material, one which holds promise as a renewable resource for the production of many chemicals and materials needed or desired by humankind. Making the cellulose in the lignocellulosic matrix more accessible to conversion interests is a prerequisite accomplishment for its expanded utility, and resolution of this problem has to be one of the more important and challenging biomass research endeavors today.

Stillage Processing for Nutrient Recovery and Pollution Control

Stillage is the liquid residue from the fermentation of grain or other feedstocks in the production of ethyl alcohol for fuel or beverages. The feasibility of producing ethyl alcohol fuel from agricultural commodities depends heavily upon the potential to capture and utilize nutrients from the stillage for animal feedstuffs or human foods. If the nation were to replace 10 percent of its gasoline consumption with ethanol, the liquid stillage from ethanol fuel production would constitute a biochemical oxygen demand (BOD_5) load equivalent to all its raw domestic sewage. Therefore, both nutrient recovery and water pollution control must be major goals of the emerging fuel-alcohol industry.

Previous Research

Stillage is recovered from the base of the beer distillation column at a temperature of about 95°C. Whole stillage from grain fermentation and distillation contains about 5 to 10 percent solids (Wu, Sexson, and Wall, 1980). It is composed primarily of unconverted starches and sugars, protein, minerals, and various fermentation products such as yeast (Barney and Chang, 1980). Stillage volume usually ranges from

10 to 13 times the volume of anhydrous alcohol (Barney and Chang, 1980). The biochemical oxygen demand (BOD_5) of stillage liquids is about two orders of magnitude higher than that of typical raw domestic sewage (Sweeten et al., 1982b).

Research on the nutritional value of stillage has dealt mostly with the dehydrated by-products of the beverage alcohol industry (Ward and Matsushima, 1980), such as distillers' dried grains with solubles (DDGS), which are easy to store and are readily marketable commodities. However, dehydration of stillage does not appear to be feasible for small-scale plants (National Research Council, 1981). Therefore, small-scale alcohol plants are faced with handling wet stillage by-products that are difficult to store, may require further processing, and have unproven nutritional and market values (National Research Council, 1981).

Nutrient recovery processes may reduce the concentration of water pollutants in stillage and return revenue to the ethanol plant. Bryan, Newton, and Johnson (1981b) concentrated suspended solids in corn stillage from a commercial distillery using a vibrating screen (500-μm aperture) and an inclined screw-press cylindrical screen (580-μm hole size). About half the suspended solids remained in the stillage effluents. More than half the weight of suspended solids in effluent was associated with particles finer than 45 μm.

Wu, Sexson, and Wall (1980) separated stillage from a fuel-alcohol corn-fermentation plant into supernatant and residue fractions by screening and centrifugation. The supernatant primarily contained degraded materials with molecular weights below 10,000. The supernatant accounted for 20 percent of the dry weight and 20 percent of the total nitrogen. The screened residue and centrifuge cake were combined into a base-of-still residue that contained 80 percent of the dry weight and 80 percent of the total nitrogen. Protein in the base-of-still residue had lowered solubility as compared with ground corn. Average crude protein contents ranged from 24 percent for residue removed by a 20-mesh screen to 42 percent for centrifuge cake (10,400 times gravity) recovered from screened stillage liquids.

Lee et al. (1981) utilized ultrafiltration (UF) and reverse osmosis (RO) to concentrate feedstock sugars prior to fermentation, to concentrate nutrients for distillers' dried grains (DDG) from the beer, and to produce a high-molecular-weight DDG concentrate from the stillage. Permeates from ultrafiltration were treated further by electrodialysis to produce high-quality water for recycling. Ultrafiltration retained the following proportion of sugar cane stillage constituents in the concentrate: total solids, 27 percent; suspended solids, 100 percent; total Kjeldahl nitrogen (TKN), 79 percent; COD, 99 percent; ash, 21 percent. The concentrate solution from reverse osmosis of UF permeate retained 100 percent of the TKN and suspended solids, 96 percent of the total solids, 89 percent of the ash, and 71 percent of the COD.

Corn thin stillage collected below a 100-mesh screen was ultrafiltered to yield 90 percent water recovery (Lee et al., 1981). The UF concentrate contained the following percentages of thin stillage constituents: 46 percent of total solids, 100 percent of suspended solids, 6.1 percent of ash, 56 percent of COD, and 71 percent of TKN. The UF permeate was subjected to reverse osmosis, which removed 97 to 99 percent of the total solids, total dissolved solids, and TKN; 82 to 99 percent of the ash; and 94 percent of the COD. The value of the animal feed derived from evaporating the UF and RO concentrates would be 65 percent higher than the cost of fuel for such evaporation (Lee et al., 1981).

Anaerobic digestion of stillage from both grain and sugar distilleries has been studied on a pilot-plant scale (Seely and Spindler, 1981; Samuels, 1980). Results showed high rates of conversion of the stillage volatile solids into biogas.

Sugar cane stillage from a rum distillery in Puerto Rico contained 8.55 percent total solids and was utilized as liquid fertilizer (Samuels, 1980). Potassium content (8.5 percent dry basis) was much higher than nitrogen or phosphorus (0.88 and 0.35 percent, respectively). Nitrogen and phosphorus were mostly in the form of colloidal organic matter. Stillage also contained important micronutrients such as sulfur (3.5 percent), calcium (2.3 percent), and magnesium (1.8 percent). Stillage pH was 4.6. Excessive application rates retarded maturation of sugar cane and decreased the quality of juice. Application rates up to 7.7 m^3/ha did not adversely affect cane quality.

Stillage Research for Texas Producers

An intensive program of research and demonstration on ethyl alcohol production using commodities grown in Texas included several objectives related to stillage processing and evaluation:

—to characterize stillage from fermented corn, grain sorghum, and other feedstocks with respect to nutrient and water pollution parameters;

—to determine low-energy processing systems for recovering nutrients from stillage;
—to develop stillage processing and treatment systems for reducing water pollution potential;
—to determine the palatability, digestibility, and nutritional value of stillage fractions as animal feedstuffs.

Detailed results of these experiments were reported in Sweeten et al. (1983a), Sweeten et al. (1983b), and Schelling et al. (1983).

Whole stillage was collected from the base of the beer distillation column and pumped into the stillage holding tank. Stillage temperature typically was 97°C at the base of the still and 82°C in the stillage holding tank. The volume of stillage per batch ranged from 1,350 to 1,600 L and averaged 14 L per liter of 184-proof ethanol yield. Whole stillage had a tendency to stratify in the holding tank.

Each batch of whole stillage was separated into wet solids and thin stillage (or liquid) fractions, using a Brown Model 3600 screw dewatering press (Brown International Corporation, Covina, California) with a screen area of 0.11 m^2 and a screen aperture of 0.315 mm. The press was driven by a 1.5-kW electric motor and was operated at pressures of 180 to 220 kPa. Mean apparent flow rate through the dewatering press ranged from 9 to 21 L/min and averaged 15 L/min.

The amount and chemical characteristics of stillage were determined in order to ascertain the potential for both animal nutrient recovery and water pollution. Composite samples of wet solids and liquid fractions were analyzed for the following parameters: moisture, ash, TKN, total sugars, starch, crude fiber, total volatile solids, filterable and nonfilterable solids, filterable volatile solids, settleable solids, chemical and biochemical oxygen demand, pH, and electrical conductivity. Concentrations of potassium, calcium, magnesium, sodium, copper, manganese, zinc, and ammonia were determined as were those of nine essential and seven nonessential amino acids.

Stillage Nutrient Concentrations. The wet pressed solids fraction of corn stillage contained an average of 31.3 percent dry matter (Table 4.7). It had very low ash (fixed solids) content and was also low in total sugars. The nitrogen concentration of wet pressed solids was 150 percent higher than the nitrogen content of the corn grain feedstock. The starch content of wet pressed solids ranged from 21.5 to 55.6 percent and averaged 32.8 percent, dry matter basis. The liquid fraction (thin stillage) from corn had an average dry matter content of

TABLE 4.7 Nutrient Concentrations in Stillage Fractions from Dewatering Press

Nutrient Parameter	Corn Stillage[(a)] (20 batches)		Grain Sorghum Stillage[(a)] (25 batches)	
	Wet Solids	Liquids	Wet Solids	Liquids
Dry matter, (DM), %	31.3 ± 0.5	3.7 ± 0.1	31.8 ± 0.4	5.8 ± 0.4
Ash, % DM	1.3 ± 0.1	7.4 ± 0.5	2.5 ± 0.4	6.5 ± 0.7
Nitrogen[(b)] % DM	5.2 ± 0.3	3.9 ± 0.2	5.5 ± 0.2	3.0 ± 0.2
Total sugars, % DM	2.2 ± 0.4	26.2 ± 2.4	5.9 ± 0.9	44.7 ± 3.9
Starch, % DM	32.8 ± 2.0	15.1 ± 1.2	28.9 ± 1.0	17.5 ± 1.1
Crude fiber, % DM	12.6[(c)]	4.9[(c)]	12.2 ± 0.6	2.6 ± 0.3

(a) Mean value ± one standard deviation.
(b) Total Kjeldahl nitrogen (TKN).
(c) Mean of three batches only.

3.7 percent, containing 7.4 percent ash, and 3.9 percent TKN. Starch and total sugars varied widely.

For grain sorghum stillage, the wet pressed solids fraction from 25 batches contained an average of 31.8 percent dry matter, which, in turn, comprised 2.5 percent ash, 5.5 percent TKN, 5.9 percent total sugars, and 28.9 percent starch (Table 4.7). The concentrations of dry matter, nitrogen, and crude fiber had small standard deviations from the mean values. The thin stillage fraction from grain sorghum fermentation contained an average of 5.8 percent dry matter, of which total sugars, starch, and crude fiber accounted for 44.7 percent, 17.5 percent, and 2.6 percent of the dry matter, respectively. The liquid fraction of grain sorghum stillage contained more dry matter, less nitrogen, and more total sugars than did the liquid fraction of corn stillage.

Mass Yields from the Dewatering Press. Dry matter and nutrient mass recovered in stillage as compared to grain feedstocks were computed (Table 4.8). Corn stillage contained 35 percent of the feedstock on the average. The wet pressed solids fraction contained less than half (46 percent) of the dry matter of whole stillage. Essentially all of the

TABLE 4.8 Dry Matter and Nutrient Yields per Batch of Stillage (dry matter basis)

	Feedstock (kg)	Recovered in Stillage			
		Pressed Solids (kg)	Liquid Fraction (kg)	Total (kg)	% of Feed-stock
Corn					
Dry matter	246.4 ± 5.3	39.8 ± 2.5	46.4 ± 1.3	86.2 ± 2.5	35
TKN	3.7 ± 0.2	2.0 ± 0.2	1.8 ± 0.1	3.8 ± 0.2	103
Total sugars	5.2 ± 0.2	0.7 ± 0.1	12.1 ± 1.5	12.8 ± 1.5	246
Starch	179.1 ± 5.3	12.2 ± 0.8	6.5 ± 0.6	18.6 ± 1.0	10
Grain sorghum					
Dry matter	263.0 ± 3.9	49.2 ± 2.8	67.4 ± 7.6	116.6 ± 10.0	44
TKN	4.2 ± 0.2	2.7 ± 0.1	1.9 ± 0.2	4.6 ± 0.3	107
Total sugars	3.2 ± 0.1	2.4 ± 0.5	28.5 ± 5.7	30.9 ± 6.3	966
Starch	118.7 ± 4.1	14.2 ± 1.2	11.2 ± 1.2	25.4 ± 2.1	22

Note: Means and standard deviations for 11 batches corn and 13 batches grain sorghum.

feedstock nitrogen was present in whole stillage, with 53 percent recovered in wet pressed solids. Only 10 percent of the starch in corn was recovered in stillage, and two-thirds of that in the wet pressed solids. Total sugars were 2.5 times greater in the stillage than in feedstock, with 95 percent of the stillage sugars present in the liquid fraction.

Grain sorghum stillage contained 44 percent of the original 263 kg of feedstock dry matter, with 42 percent of that dry matter recovered by the dewatering press as wet pressed solids. Nitrogen in the grain was completely accounted for in the stillage; 60 percent was recovered in pressed solids. Only 14 percent of the grain starch remained in the stillage, more than half captured in the pressed solids. Total sugars increased more than ninefold, but less than one-tenth of the stillage total sugars was recovered in the pressed solids fraction.

Amino Acid and Mineral Content. Amino acid content increased in both the wet pressed solids and liquid fractions, as compared to the original feedstock (Tables 4.9 and 4.10). The wet solids fraction contained most of the amino acids following fermentation; however, the

TABLE 4.9 Amino Acid and Mineral Composition of Corn and Fermentation By-products

	Corn Grain	Pressed Solids	Stillage Liquids
Dry matter (%)	86.5 ± 0.5	32.6 ± 0.0	3.7 ± 0.9
	Essential Amino Acids (mg/g dry matter)		
Arginine	5.0 ± 0.4	12.0 ± 1.0	9.8 ± 1.7
Histidine	2.7 ± 0.2	7.4 ± 0.2	5.5 ± 0.8
Isoleucine	3.4 ± 0.6	9.9 ± 0.4	7.8 ± 0.3
Leucine	11.6 ± 0.1	37.5 ± 1.3	22.5 ± 0.7
Lysine	3.1 ± 0.5	6.6 ± 0.5	8.0 ± 1.1
Methionine	2.2 ± 0.04	6.3 ± 0.6	5.5 ± 0.7
Phenylalanine	4.8 ± 0.4	14.7 ± 0.8	9.5 ± 0.7
Threonine	3.5 ± 0.5	10.5 ± 1.1	7.5 ± 0.4
Valine	7.3 ± 0.7	21.2 ± 2.1	15.6 ± 1.0
	Nonessential Amino Acids (mg/g dry matter)		
Alanine	6.4 ± 0.2	19.2 ± 1.5	13.3 ± 1.2
Aspartic acid	6.6 ± 1.0	16.2 ± 2.0	14.4 ± 1.7
Glutamic acid	18.2 ± 1.3	52.9 ± 3.6	35.5 ± 3.1
Glycine	3.4 ± 0.2	8.7 ± 0.8	7.7 ± 0.9
Proline	8.2 ± 0.4	22.3 ± 6.7	16.9 ± 2.0
Serine	4.3 ± 0.06	11.8 ± 1.0	9.5 ± 0.6
Tyrosine	4.0 ± 0.3	12.5 ± 0.7	8.4 ± 0.5
Ammonia	2.9 ± 0.1	8.7 ± 0.2	6.9 ± 0.9
	Minerals		
Ash (%)	1.3 ± 0.07	1.4 ± 0.3	5.2 ± 0.8
Calcium (%)	0.013 ± 0.004	0.036 ± 0.015	0.174 ± 0.068
Potassium (%)	0.267 ± 0.016	0.104 ± 0.005	0.617 ± 0.094
Magnesium (%)	0.144 ± 0.001	0.082 ± 0.009	0.400 ± 0.090
Sodium (%)	0.094 ± 0.016	0.110 ± 0.005	1.76 ± 1.10
Copper (ppm)	1.5 ± 1.0	10.4 ± 4.8	23.8 ± 2.4
Manganese (ppm)	4.6 ± 0.6	4.3 ± 2.2	23.9 ± 5.1
Zinc (ppm)	62.5 ± 6.6	68.0 ± 1.5	172.8 ± 0.1

Note: Means ± SD; mineral and ash concentrations expressed on a dry matter basis.

TABLE 4.10 Amino Acid and Mineral Composition of Grain Sorghum and Fermentation By-products

	Sorghum Grain	Pressed Solids	Stillage Liquids
Dry matter (%)	88.5 ± 0.1	31.1 ± 0.2	9.0 ± 1.0
	Essential Amino Acids (mg/g dry matter)		
Arginine	4.2 ± 0.4	10.3 ± 0.05	6.2 ± 0.5
Histidine	2.3 ± 0.2	6.7 ± 0.2	3.7 ± 0.6
Isoleucine	3.9 ± 0.4	12.3 ± 0.3	5.9 ± 0.2
Leucine	13.0 ± 1.2	43.1 ± 0.1	16.2 ± 0.5
Lysine	2.4 ± 0.01	5.2 ± 0.03	5.4 ± 0.5
Methionine	1.9 ± 0.3	6.2 ± 0.4	3.2 ± 0.1
Phenylalanine	5.1 ± 0.4	16.8 ± 0.3	7.0 ± 0.03
Threonine	3.5 ± 0.1	9.7 ± 0.2	5.8 ± 0.6
Valine	8.3 ± 0.6	24.8 ± 0.9	11.9 ± 0.1
	Nonessential Amino Acids (mg/g dry matter)		
Alanine	8.4 ± 0.6	26.7 ± 1.0	11.0 ± 0.5
Aspartic acid	6.0 ± 0.4	18.6 ± 1.1	10.7 ± 0.8
Glutamic acid	20.4 ± 1.4	65.4 ± 3.3	24.9 ± 1.2
Glycine	3.1 ± 0.1	8.2 ± 0.1	5.2 ± 0.4
Proline	7.0 ± 0.1	27.9 ±	9.9 ± 0.4
Serine	3.9 ± 0.3	12.4 ± 1.2	5.7 ± 0.6
Tyrosine	4.1 ± 0.4	13.9 ± 0.4	5.8 ± 0.3
Ammonia	3.4 ± 0.2	11.1 ± 0.01	4.7 ± 0.1
	Minerals		
Ash (%)	1.6 ± 0.2	2.5 ± 0.4	8.6 ± 0.1
Calcium (%)	0.033 ± 0.004	0.042 ± 0.014	0.135 ± 0.002
Potassium (%)	0.302 ± 0.008	0.061 ± 0.039	0.346 ± 0.126
Magnesium (%)	0.183 ± 0.006	0.116 ± 0.007	0.501 ± 0.030
Sodium (%)	0.008 ± 0.014	0.394 ± 0.127	4.61 ± 0.80
Copper (ppm)	2.5 ± 0.0	16.8 ± 14.6	10.1 ± 1.2
Manganese (ppm)	6.5 ± 1.8	4.4 ± 0.4	26.5 ± 1.6
Zinc (ppm)	86.0 ± 16.8	101.1 ± 2.3	222.5 ± 2.7

Note: Means ± SD; mineral and ash concentrations expressed on a dry matter basis.

amino acid profile of the stillage liquid fraction indicated that it has a somewhat better nutritional balance of these proteins than the wet solids fraction. The amino acid content of grain sorghum stillage fractions showed a pattern very similar to that of corn.

The protein in stillage appeared to be of more value to ruminants than that of the grain feedstock because stillage protein bypasses the rumen. The importance of bypass protein in ruminants was reviewed by Kempton, Nolan, and Leng (1977), who pointed out the potential advantage of having a significant portion of the dietary protein pass through the rumen to be digested in the small intestine. The natural resistance of stillage protein to breakdown in the rumen may be advantageous for ruminant feeds.

There was considerable difference in the distribution of individual mineral elements in the two stillage fractions. Potassium, sodium, and magnesium were leached from the solid fraction (Tables 4.9 and 4.10). The other trace minerals had a tendency to maintain or significantly increase their concentrations in the pressed solids fraction relative to the feedstock concentrations.

Digestibility by Ruminants. Cattle-feeding studies were conducted with wet pressed solids from sorghum grain to determine the digestibility of each nutrient class. This involved 30 animal observations (ten steers × three treatments). Each steer weighed 300 kg initially. Results indicated that overall dry matter digestibility was 76.9 percent. This value was determined by regressing graded feeding levels of the stillage feed to a 100 percent intake in order to correct for the digestibility of the other components in the diet. Because most of the highly digestible starch is removed during fermentation, the digestibility of the wet pressed solids in these cattle was less than that of sorghum grain. The value does represent a satisfactory digestibility, however, for the combined energy and protein content of the feed.

Water Pollutants in Thin Stillage. The dewatering press liquid fraction, or thin stillage, contained very high concentrations of certain water pollutant parameters (Table 4.11). This liquid effluent was characterized by high solids concentration, low pH, high oxygen demand, and moderate salinity levels. The total solids concentration in dewatering press liquids averaged 37,000 ppm for corn stillage (15 batches) and 60,000 ppm for grain sorghum stillage (28 batches). Grain sorghum

TABLE 4.11 Water Pollutant Concentrations for Liquid Fraction (Thin Stillage) from Dewatering Press

Constituent	Corn	Grain Sorghum
Total solids (ppm)	37,000 ± 2,200	60,000 ± 4,000
Total volatile solids (ppm)	34,300 ± 2,200	56,600 ± 3,800
Filterable (dissolved) solids (ppm)	18,500 ± 1,700	38,200 ± 3,200
Nonfilterable solids, suspended (ppm)	18,400 ± 900	22,600 ± 1,200
Chemical oxygen demand (ppm)	62,800 ± 3,000	92,700 ± 5,400
Biochemical oxygen demand (ppm)	28,700 ± 1,000	27,300[a]
pH	4.24 ± 0.03	4.01 ± 0.05
Electrical conductivity (μmhos/cm)	2,430 ± 90	3,500 ± 240
Settled solids interface height (%)		65 ± 5

Note: Means ± standard deviations.
(a) Only one sample was tested.

stillage apparently contained a higher proportion of fine suspended and dissolved organic matter. Specific gravity of 14 batches of grain sorghum stillage dewatering press effluent was 1.021 ± 0.003. Thin stillage from both corn and grain sorghum had low pH, moderate salinity, and high chemical oxygen demand (Table 4.11). Biochemical oxygen demand (BOD_5) for corn stillage ranged from 25,500 to 31,200 ppm and averaged 28,700 ppm; and, there was little apparent difference between grain sorghum and corn stillage liquids. By contrast, the U.S. Environmental Protection Agency (EPA) normally requires less than 20 ppm BOD_5 and 20 ppm total suspended solids (or nonfilterable solids) in plant wastewater before they can be discharged.

In comparing the water pollutant concentrations in thin stillage in Table 4.11 with the EPA effluent guidelines and new source performance standards for corn wet-milling plants, a related industry, it is evident that major wastewater control efforts will have to be made at ethyl alcohol plants. The EPA effluent limitations for new corn wet-milling plants are 0.36 kg BOD_5 and 0.45 kg total suspended (or non-filterable) solids per 1,000 kg of corn processed. The corn thin stillage data in Table 4.11 indicate that an ethanol plant produces 130 kg BOD_5 and 83 kg total suspended solids per 1,000 kg of corn used, assuming 4.5 L of thin stillage per kg corn feedstock. Thus, a pollutant reduction of 99.99 percent and 99.45 percent for BOD_5 and total suspended solids, respectively, is needed in the treatment of thin stillage as a wastewater. Moreover, the EPA effluent limitations stipulate a pH limit of 6.0 to 9.0 for corn wet milling and many other processes as well. Thus, the acidity (pH = 4.24) of corn and grain sorghum stillage will also have to be addressed by abatement technology.

Centrifugation. Thin stillage from several batches of corn and grain sorghum were centrifuged to further assess the nutrient recovery possibilities. In some cases, centrifuge supernatant was treated further by ultrafiltration, reverse osmosis, or a combination of these two processes. In addition to constituent concentration differences before and after processing, mass balances were calculated for dry matter, nitrogen, total sugar, starch, and ash. For grain sorghum thin stillage, centrifugation at 6000 × G lowered the total solids concentration by 28 percent, COD by 43 percent, and TKN by 65 percent (Table 4.12). Neutral detergent fiber (NDF, a measure of cellulose, hemicellulose, and lignin) was reduced by 97 percent. Centrifuge cake accounted for 18.8 percent of the thin stillage processed, 34 percent of the dry matter, and 81 percent of the nitrogen on a mass basis.

Reverse Osmosis of Centrifuge Supernatant. Reverse osmosis (RO) lowered the total solids and volatile solids (VS) concentrations of centrifuge supernatant by 92 and 93 percent to values of 3,630 and 2,780 ppm, respectively, in the RO permeate (Table 4.12). The COD concentration was reduced by 82 percent to 9,350 ppm in RO permeate, and salts (electrical conductivity) were lowered by 62 percent. Essentially 100 percent of the starch and fiber and 92 percent of the total sugars were removed from the wastewater by RO. Nitrogen removal occurred through removal of total solids, but the TKN concentration as a percent of dry matter was unchanged.

The volume of RO concentrate was only 10.8 percent of the centrifuge supernatant processed. The RO concentrate was enriched in terms of total sugars, starch, and volatile solids. This material contained 27.3 percent total solids. Energy requirements and cost for de-

TABLE 4.12 Nutrients and Pollutants in Thin Stillage after Centrifugation and Reverse Osmosis

		Stage I Centrifugation			Stage II Reverse Osmosis			
Constituent	Thin Stillage (inflow)	Cake	Supernatant	% Reduction[(a)]	Concentration	Permeate	% Reduction	Overall Reduction (I & II) %
Total solids (ppm)	64,100 ±11,900	134,000 ±14,500	46,500 ±12,200	28	272,900 ±88,500	3,630 ±1,600	92	94
Volatile solids (ppm)	60,600 ±12,300	129,600 ±14,500	43,200 ±12,600	29	249,200 ±87,800	2,780 ±1,370	93	95
Chemical oxygen demand (ppm)	93,400 ±10,200	—	53,400 ±18,600	43	399,200 ±147,400	9,350 ±4,760	82	90
Electrical conductivity (μmho/cm)	2920 ±580	2030	3010 ±1160	—	9930 ±3490	1130 ±340	62	61
Starch (% d.b.)	32.0 ±11.7	24.3 ±5.4	36.5 ±11.5	—	36.3 ±11.8	0.0088	100	100
Total Kjeldahl nitrogen (% d.b.)	1.70 ±0.50	3.49 ±0.91	0.59 ±0.36	65	0.70 ±0.43	0.66 ±0.59	—	61
Total sugars (% d.b.)	39.6 ±13.4	19.0 ±7.0	53.6 ±14.4	—	51.7 ±15.8	3.6 ±4.4	93	91
Neutral detergent fiber (% d.b.)	24.1 ±6.5	38.7 ±4.5	0.62 ±0.51	97	0.31 ±0.29	0.0	100	100

Note: Mean values ± standard deviation for four test runs.

(a) Concentration of permeate relative to liquid feed from preceding treatment stage.

hydration and recovery of animal feed from this concentrate may not be excessive. It appears that RO concentrate could be a useful feedstock for anaerobic digestion to produce biogas or methane, or could be recycled for ethanol production.

Ultrafiltration of Centrifuge Supernatant. The ultrafiltration (UF) concentrate contained 16 percent of the mass of centrifuge supernatant processed. Its total solids concentration was only 5.7 percent. The UF unit removed substantially more nitrogen (TKN) than the RO unit but was not nearly as efficient in the removal of total solids, volatile solids, COD, or salt (Table 4.13). The total sugar content on a dry matter basis was higher in the UF permeate than in the UF concentrate. The UF concentrate would have less potential than RO concentrate as an ethanol or anaerobic digester feedstock.

Reverse Osmosis of Ultrafiltration Permeate. Ultrafiltration permeate (115 kg) from one stillage batch was fed into the RO unit, yielding large concentration reductions for all parameters except sugar and ash (Table 4.13). The resulting concentrate had only 10% of the mass of the UF permeate feed, and it was relatively high in total sugar and starch. The RO permeate had less salt than the tap water used to produce the ethyl alcohol and stillage (EC values of 270 vs. 770 μmho/cm).

This experiment revealed that the overall reduction in pollutant concentration for a three-stage thin stillage treatment system consisting of centrifugation, UF, and RO may be as large as 99 percent removal of total volatile solids, 96 percent removal of COD, and 100 percent removal of starch, nitrogen, and fiber.

Particle-Size Distribution. Particle-size distribution was determined on stillage liquid fractions from two test runs, both before and after processing with the screw press, centrifuges, ultrafiltration, and reverse osmosis. A Coulter Counter Model TAII (Coulter Electronics, Inc., Hialeah, Florida) was used to determine the percent of particles by volume and population of a specified size within the particle-size range 6.35 to 400 μm. Wet sieve analysis was conducted to determine the distribution of particles in the 0 to 1,130 μm range.

The processing methods discussed above caused profound changes in particle-size distributions of stillage materials. Wet sieve analysis of whole stillage and dewatering press liquids revealed a radical shift in predominant particle sizes. Coarse particles larger than 500 μm ac-

TABLE 4.13 Nutrients and Pollutants in Thin Stillage after Centrifugation, Ultrafiltration, and Reverse Osmosis

Constituent	Thin Stillage (inflow)	Stage I Centrifugation			Stage II Ultrafiltration			Stage III Reverse Osmosis (ultrafiltration permeate)			Overall
		Cake	Supernatant	% Reduction[(a)]	Concentrate	Permeate	% Reduction	Concentration	Permeate	% Reduction	Reduction (I, II, & III) %
Total solids (ppm)	64,100 ±11,900	134,000 ±14,500	46,500 ±12,200	28	56,600	30,700	34	208,700	790	97	99
Volatile solids (ppm)	60,600 ±12,300	129,600 ±14,500	43,200 ±12,600	29	52,600	26,400	39	192,600	450	98	99
Chemical oxygen demand (ppm)	93,400 ±10,200	—	53,400 ±18,600	43	63,600	34,300	36	251,300	4,140	88	96
Electrical conductivity (μmho/cm)	2,920 ±580	2,030	3,010 ±1,160	—	2,600	2,590	14	6,600	270	90	91
Starch (% d.b.)	32.0 ±11.7	24.3 ±5.4	36.5 ±11.5	—	44.5	39.3	11	35.3	0.0	100	100
Total Kjeldahl nitrogen (% d.b.)	1.70 ±0.50	3.49 ±0.91	0.59 ±0.36	65	0.81	0.27	55	0.21	0.0	100	100
Total sugars (% d.b.)	39.6 ±13.4	19.0 ±7.0	53.6 ±14.4	—	34.6	71.2	—	72.6	38.0	47	4
Neutral detergent fiber (% d.b.)	24.1 ±6.5	38.7 ±4.5	0.62 ±0.51	97	2.3	0	100	0.39	0	100	100
Ash (% d.b.)	5.8 ±1.9	3.3 ±0.5	8.0 ±3.4	—	6.9	10.2	—	7.7	43.0	—	(−740)

Note: Mean values ± standard deviation for four test runs.
(a) Concentration of permeate relative to liquid feed from preceding treatment stage.

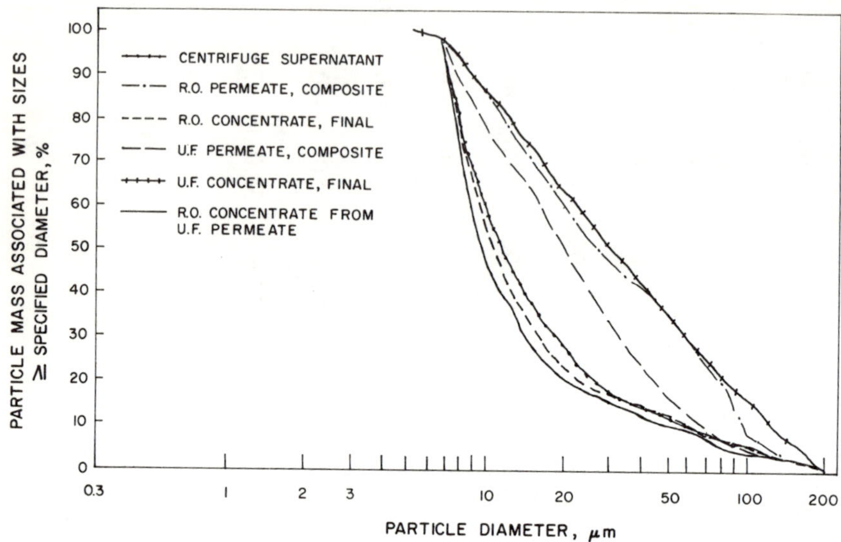

FIG. 4.12. Changes in particle-size distributions in processing thin stillage by centrifugation, ultrafiltration, and reverse osmosis (as determined with the Coulter Counter).

counted for 45 percent of the dry matter weight of whole stillage, while particles of 0 to 43 μm represented 38 percent of the whole stillage dry matter. There was a fairly even distribution of particles between 43 and 500 μm, which together constituted only 15.4 percent of the whole stillage dry matter.

The dewatering screw press (315-μm screen aperture) removed almost all of the particles of 419-μm diameter or larger so that together these particles constituted only 1.2 percent of the total solids. Many of the 250- to 419-μm particles are also removed. The resulting thin stillage contained a large majority (78 percent) of particles with sizes of 0 to 43 μm.

Changes in particle-size distribution that resulted from centrifugation, ultrafiltration, and reverse osmosis of stillage are shown in Figure 4.12. This graph shows an analysis from the Coulter Counter for 6.35- to 200-μm particles. Median particle sizes were as follows: thin stillage, 21.7 μm; centrifuge supernatant, 11.4 μm; RO concentrate (final), 20.0 μm; UF permeate (composite), 10.6 μm; RO concentrate from UF permeate, 9.5 μm; UF concentrate (final, 30.5 μm; RO permeate (composite), 25.8 μm; and RO permeate of UF permeate, 26.9 μm. In essence, the RO permeate and UF concentrate contained coarser particles than the thin stillage, while other processing fractions had finer particles.

Summary and Conclusions

Whole stillage contained 35 and 44 percent, respectively, of the dry matter originally present in corn and grain sorghum feedstocks to an ethyl alcohol plant. Dry matter recovery from a commercial dewatering press was 42 percent from grain sorghum stillage and 46 percent from corn stillage. Wet pressed solids from corn and grain sorghum stillage contained 31 to 32 percent dry matter, of which about 2 percent was ash, 5.2 to 5.5 percent total nitrogen, 30 percent starch, and 12 percent crude fiber. The liquid fraction, or thin stillage, from the dewatering press contained 3.7 and 6.0 percent dry matter for corn and grain sorghum, respectively. Ash content was approximately 7 percent, and total nitrogen was 3.9 percent for corn and 3.0 percent for grain sorghum.

Amino acids were concentrated two- and threefold in thin stillage and pressed solids dry matter, respectively. The dry matter digestibility of the wet pressed grain sorghum solids fraction for cattle was 76.9 percent.

Thin stillage from the ethanol plant was highly polluted wastewater, containing 28,500 ppm BOD_5, 63,000 to 93,000 ppm COD, and a pH of 4.0 to 4.2. A centrifugation removed 21 percent of the dry matter and 46 percent of the nitrogen from thin stillage, yielding a centrifuge cake containing 43 percent crude protein. Another type of centrifuge removed 34 percent of the dry matter and 81 percent of the nitrogen on a mass basis. Corresponding concentration reductions in supernatant as compared to thin stillage were 28 percent for dry matter and 65 percent for crude protein.

Reverse osmosis (RO) produced 92 percent reduction in the total solids concentration of centrifuge supernatant and 82 percent reduction in COD. Almost all the starch, sugar, and fiber were retained in the RO concentrate, which had one-tenth the mass of centrifuge supernatant. The RO concentrate may be useful for ethyl alcohol plant feedstock, biogas production through anaerobic digestion, or dehydration for animal feedstuffs.

Ultrafiltration (UF) removed more nitrogen from centrifuge supernatant than RO but was less effective in removing starch, sugars, total solids, and chemical oxygen demand. The UF permeate was processed further by RO to produce an overall reduction of 96 to 100 percent in concentrations of most water pollutant parameters and nutrients. However, the remaining wastewater still contained high concentrations of total solids and COD, and would require additional wastewater treatment or land disposal.

Particle-size distribution of whole stillage was drastically altered by the processing methods employed. The screw press removed grain sorghum stillage particles larger than 250 μm. Almost 80 percent of the particles in thin stillage were smaller than 43 μm, and 50 percent of those were smaller than 7.7 μm.

The following conclusions were drawn from the research:

1. Grain sorghum stillage from ethyl alcohol production was a rich source of animal nutrients.

2. Almost all of the animal nutrients could be harvested from the stillage stream by the following combinations of processes: (a) dewatering screw press, centrifugation, and reverse osmosis; or (b) screw press, centrifuge, ultrafiltration, and reverse osmosis.

3. Wastewater remaining after three or four stages of treatment was still highly contaminated and would require further treatment or land application.

Ethanol Use

Alcohol, particularly ethanol, has been considered for use as a fuel in internal combustion engines ever since they were developed. Samuel Morey of Oxford, New Hampshire, recommended alcohol as a fuel for the engine he patented in 1826. Sorel, Woodward, and Preston (1907) and Strong (1909) presented information on carbureting, combustion, engine design, operating conditions, and performance results for alcohol fueled engines.

Most mobile equipment used on the farm today is powered by the compression ignition (CI) engine which uses diesel fuel. However, many spark ignition (SI) engines are still being utilized in agriculture, as well as in personnel transportation, which use gasoline, natural gas, or propane as fuel. Recent research has demonstrated several methods for converting naturally aspirated and turbocharged CI and SI engines to utilize ethanol as either a fuel extender (dual-fueling) or as the primary fuel.

Compression Ignition Engines

The use of alcohol in place of, or as a supplement to, diesel fuel has occurred within the past decade. Until recently, many textbooks had flatly stated that it was impossible to use alcohol in diesel engines. While this view is now changing, there are still properties of alcohol that, when compared to diesel fuel, present problems (Adelman, 1980; Hagen, 1980). Alcohol's high octane rating is a measure of its resistance to autoignition when exposed to heat and pressure, which means it has a low cetane number. Alcohol's lack of lubricant is also a major problem for the fuel injection systems of CI engines (Adelman, 1980).

Ethanol might better be utilized as a fuel extender, rather than as a total replacement for diesel fuel. There are several methods for dual-fueling CI engines with alcohol and diesel fuels, including fumigation and emulsification.

Fumigation

The process in which ethanol and diesel fuels are introduced separately into CI engines is known as dual-fueling. Ethanol enters the air intake system between the air cleaner and intake manifold. Diesel fuel is injected into the combustion chamber and acts as a pilot to ignite the ethanol and diesel fuel. Generally, there is an increase in ignition delay in CI engines when ethanol is fumigated in droplet form. Ethanol has a relatively high heat of vaporization (842 kJ/kg). The vaporization in the combustion chamber of a stoichiometric amount of ethanol mixed with air lowers the adiabatic mixture temperature by 93°C, thus causing ignition delay until the mixture is returned to its self-ignition temperature (Adelman, 1980).

The maximum power output of a CI engine is frequently increased by dual-fueling with alcohol. CI engines have excessive air in the combustion chamber, air that is not used in the combustion process; the alcohol mixes with this excess air to increase maximum power output. The degree of power increase depends upon the method used to induct the ethanol into the air intake system.

Several methods of fumigation have been researched; they include (a) carburetion, (b) air pressure from turbocharger, (c) low-pressure intake manifold injection, and (d) alcohol in vapor form.

Carburetion. In a direct-injection CI engine described by Pringle and Mayeux (1982), aqueous ethanol is carbureted into a secondary air intake line. The carburetor is sized so that there is no significant increase in airflow resistance. The throttle and choke plates are frequently removed or held open. The flow of ethanol is controlled by a needle valve. This type of system may also be used on turbocharged engines if the airflow rate is not changed.

Pringle and Mayeux (1982) also found that when an engine is at full load, 23 percent of the diesel fuel could be replaced with 190-proof ethanol carbureted at a flow rate of 35 percent by weight. At full-load and three-quarter-load conditions, this method of dual-fueling was as effective as diesel fuel alone; however, thermal efficiency was reduced when the engine operated at one-half and one-quarter load and rated speed. Forty-six percent of the fuel energy of a tractor could be supplied by carburetion with little effect on the engine's thermal efficiency (Cruz, Rotz, and Watson, 1982).

Turbocharger Air Pressure System. Turbocharged engines provide the option of spraying intake air with ethanol through either a nozzle or an orifice (Walter and Kaufman, 1980; Sullivan, 1981). A variable-flow pump is used to control the flow of alcohol through the nozzle. The fuel flow through the orifice is controlled by pressure from a tank pressurized by the turbocharger. Pre-turbocharger alcohol fumigation is not recommended (Sullivan, 1981) because of excessive turbocharger blade wear. In one 30-hour test the blades showed ten times the wear of those in a typical tractor operated for 1,500 hours.

In general, the diesel fuel was replaced by an equivalent energy content of alcohol. The thermal efficiency was maintained or slightly increased at full load and reduced slightly at partloads. The volumetric efficiency of the engine was reduced slightly, and power was increased as much as 36 percent.

Low-Pressure Intake Manifold Injection System. A low-pressure gasoline injection system was used to spray ethanol into the manifold at the intake port of each cylinder of an engine as the valve opened for the intake stroke. The nozzles were connected to a distributor valve, which controlled the flow rate and thus insured equal flow to each nozzle. The amount of diesel fuel that the ethanol replaced varied with engine speed and load. Ethanol could supply 66 percent of the fuel at full load (Shropshire and Goering, 1982); 30 to 50 percent of the diesel fuel could be replaced at two-thirds and one-third load. Brake thermal efficiency was increased at full load but decreased at part load.

Alcohol Vapor Fumigation. A system utilizing the principles developed by Harris and Davison (1980) for vaporizing ethanol prior to fumigation has been used to supply alcohol to a diesel engine. The major components of the system include the alcohol tank, filter, pump, vaporizer, and pressure-reducing valve (Aldred, Prebeg, and Thomson, 1983). The vaporizer utilizes heat from the engine exhaust gases to produce alcohol vapor. The engine is started as a diesel engine. The pressure of the vapor is controlled by the pressure setting of the pump and the pressure-reducing valve. Alcohol vapor at atmospheric pressure is transported through the air intake system and into the combustion chamber by the vacuum produced in the manifold on the intake stroke. The alcohol-air mixture is ignited when the diesel fuel is injected. Up to 63 percent by volume of the fuel supplied at three-quarters and full load of a naturally aspirated, direct-injection, three-cylinder diesel engine was from the vaporized ethanol.

Emulsification

The use of straight ethanol as a CI engine fuel is not promising, but the use of diesel fuel for pilot ignition of an alcohol-air mixture is an approach that is likely to be successful. One of the simplest ways to use both fuels is to blend them together so that no engine modifications are required for their use. At room temperature, anhydrous ethanol and diesel fuel are miscible, but a water concentration of only 0.05 percent will cause separation.

An emulsion of diesel fuel, ethanol, and a trace of water can be formed by mechanical means or by adding surfactants. Emulsions formed by agitation remain emulsified only for a brief period. Boruff et al. (1983) found that an emulsion made from 190-proof ethanol, No. 2 diesel, and N,N-dimethylethanolamine has the desired characteristics, and the microemulsion is stable at −15.5°C (4°F).

The peak power obtained by a naturally aspirated direct injection engine without any modification from the microemulsion was 93 percent of that for diesel fuel. Thermal efficiency at maximum power was 4 to 5 percent greater for the blended fuel than for diesel fuel. Straight diesel gave better efficiency at less than maximum loads. The exhaust and coolant temperatures were lower with the blended fuel.

Spark Ignition Engines

Schrock (1980) gave an excellent summary of basic fuel properties, including those of gasoline and ethanol. The energy content of a liter of gasoline is 32.47 MJ compared to 21 MJ for ethanol. The stoichiometric air-fuel ratio is 14.7 : 1 for gasoline and 9 : 1 for ethanol. Therefore, less oxygen is required for the efficient combustion of alcohol than of gasoline. The intake air temperature will be lower for an engine burning alcohol because of alcohol's higher heat of vaporization. These factors make it necessary to modify the gasoline carburetor.

Several methods have been tried to convert gasoline-fueled SI engines to permit operation with alcohol, the most common method being modification of the gasoline carburetor. This modification is necessary in order for the engine to develop the same power as with gasoline; it includes enlarging carburetor jet sizes to increase the fuel flow and replacing parts in the carburetor that alcohol would destroy.

Other modifications include (a) increasing the heat to the intake manifold, (b) advancing the timing, (c) increasing the compression ratio, and (d) providing starting aids for temperatures below 40°C (Braun and Stephenson, 1982).

Bashford and Sullivan (1981) converted a Chrysler 3.69L slant-block, six-cylinder engine to operate on 190-proof alcohol. Initially, the temperature in the intake manifold was 13°C, which indicates that complete vaporization of the fuel was not occurring. A heat shield was then constructed around the intake and exhaust manifolds, and a heat exchanger was used to heat the alcohol before it entered the carburetor. These two actions increased the temperature in the intake manifold to 24°C. Maximum power was 39 kW with gasoline as a fuel, and 36 kW with 190-proof alcohol. Brake-specific fuel consumption was higher for alcohol than for gasoline. The thermal efficiencies of the two fuels were approximately equal over the power range tested.

A gasoline tractor, modified for operation on 180-proof ethanol by Bassett and Chisholm (1980), had a decrease in maximum power of 19 percent. The tractor engine operated at 11 percent higher thermal efficiency on ethanol.

The high octane rating of ethanol suggests its use as a fuel in engines with a higher compression ratio than that of present-day SI engines. Of the many variables affecting the efficiency of an engine, the compression ratio has a significant and well-documented effect on fuel economy. Increasing the compression ratio of current SI engines from 8 to 11 or 12 should increase fuel economy by approximately 20 percent (Most and Longwell, 1980).

Anhydrous Ethanol and Gasohol

Anhydrous ethanol not only has a high octane rating but mixes with gasoline, provided there is no water present. A mixture of 90 percent gasoline and 10 percent anhydrous ethanol, called gasohol, can be used as fuel in most SI engines without changes or adjustments to the engine carburetor or timing. Gasohol was promoted during the 1979 fuel crisis; however, the name is not presently being used very extensively. In its place, terms such as "unleaded gasoline with ethanol" are being employed.

Octane Enhancement

Over the past few years, some of the cheaper and more effective chemicals and compounds once used to improve the octane rating of unleaded gasoline have been removed from the market. Alcohol is now being increasingly used instead. When alcohol is added to gasoline, the increase in octane rating is affected by the octane value of the unblended gasoline, with low-octane fuels showing more pronounced increases than high-octane fuels for the same fraction of alcohol addition (Bolt, 1980). One major advantage of using alcohol to boost octane and to extend diminishing gasoline supplies is that the gasoline-alcohol mixture is compatible with existing SI engines without any alteration. Also, using alcohol as an octane enhancer for unleaded gasoline allows distribution through present supply systems without any major changes.

Potter (1983) estimated that ethanol blends of unleaded premium would reach 4 percent of the gasoline market in 1983. In 1982, unleaded gasoline accounted for about 53 percent of total gasoline sales of some 360 billion liters (Simmons, 1983). There is still room for considerable growth in the ethanol industry.

Summary

Fermentation ethanol production from grains has increased dramatically in the U.S. in recent years. The traditional process of converting starch to ethanol is still the most widely used, although new processes are being developed and implemented. The basic process consists of converting starch to sugar by enzyme hydrolysis. Yeasts are then added to ferment the sugars into ethanol. Following fermentation, the ethanol is removed from the beer by conventional and azeotropic distillation to obtain anhydrous ethanol, which is then blended with gasoline and used for automotive fuel.

Traditionally, the starch conversion and fermentation processes have been carried out in a batch mode, but continuous processes have been developed and are being implemented for both of these operations.

Fractional distillation is used almost exclusively for separating 190-proof ethanol from the beer, and no alternative appears promising for the near future. Ethanol dehydration technology, for removing the remaining water from the ethanol, has seen numerous innovations, some with substantially lower energy requirements than conventional azeotropic distillation. Molecular sieve dehydration systems are currently being employed in some commercial operations. Other promising techniques, such as solvent extraction and starch or cellulose adsorption, are in either the laboratory or pilot-plant stage.

Much research effort is also being focused on the development of alternative feedstocks for ethanol production. Corn is by far the most common ethanol feedstock in the U.S., primarily because it is widely available, and the grain-processing technology is well developed. Other feedstocks such as sweet sorghum show potential for increased ethanol production at lower cost; however, further development of handling and processing systems is required before the full potential of some of these feedstocks is realized. There is also a need for genetic improvement of crops to enhance their ability to serve as feedstocks for alcohol production.

One area of feedstock research receiving considerable effort is cellulose conversion. Cellulose is an extremely abundant and potentially low-cost ethanol feedstock. Pure cellulose can readily be hydrolyzed to sugars with acids or enzymes. The major problem in utilizing cellulose is extracting it from lignocellulose, the form in which it normally occurs in nature; the processes that can be used to remove cellulose from the lignocellulose complex also tend to degrade the glucose formed, making it unavailable for fermentation to ethanol. Physical and chemical processes being developed have shown promise in laboratory work, however, and it appears to be only a matter of time before the problem is solved.

Stillage, the portion of the beer remaining after the ethanol has been stripped out, is a major by-product of the fermentation ethanol industry. It can be a valuable commodity or an extremely potent pollutant, depending on the final processing. In large grain ethanol plants it has traditionally been dehydrated and sold as distillers' dried grains for animal feed. Dehydration is an energy-intensive process and not generally feasible for small plants.

Minimum processing normally involves the separation of whole stillage into wet solids and thin stillage by means of a screw press or vibrating screen. The wet solids are high in crude protein and have a better amino acid balance than the original grain. Overall digestibility is about 77 percent in cattle.

Thin stillage, when considered as a pollutant, contains about 4 to 6 percent total solids and a COD ranging from 60,000 to 90,000 ppm. Treatment with centrifugation, ultrafiltration, and reverse osmosis can reduce the solids by 99 percent and the COD by 96 percent.

Currently, most anhydrous ethanol produced is used as an octane enhancer and/or gasoline extender in the U.S. Considerable research is being directed toward utilization of hydrous ethanol in both spark and compression ignition engines.

In compression ignition engines, ethanol is used to replace only a portion of the diesel fuel, and the engine is operated in a dual-fuel mode. Either the ethanol is mixed with the intake air prior to entering the combustion chamber, or an emulsion of ethanol and diesel is formed. Fueling the engine with an ethanol/diesel emulsion requires no engine modification.

Hydrous ethanol can successfully replace gasoline in spark ignition engines, although some engine modification is usually required. Most of the technology that involves fueling engines with hydrous ethanol is still in the testing phase and not yet available commercially.

References

Adelman, H. (1980). Alcohols in diesel engines—a review. In *Alcohols as motor fuels*, 267–75. Progress in Technology Series, No. 19. Warrendale, Pa.: Society of Automotive Engineers.

Ad-Pro Industries, Inc. (undated), Houston, Tex. Information letter and unit diagram.

Aldred, W. H., M. J. Prebeg, and W. A. Thomson (1983). *Retrofitting methane fueled SI engine for alcohol fuel*. ASAE paper 83-1062. St. Joseph, Mich.: American Society of Agricultural Engineers.

Allen, A. S. (n.d.). *Full-scale continuous fermentation for the commercial production of fuel ethanol.* Houston, Tex.: Bohler Bros. of America, Inc./Vogelbusch Division.

Badger, P. C., R. S. Pile, D. W. Burch, D. A. Mays, and J. W. Lewis (1982). *TVA/DOE integrated on-farm alcohol production system progress report, Phase 2: October 1981–February 1982.* Circular Z-134. Muscle Shoals, Ala.: Tennessee Valley Authority, NFDC.

Barney, W. K., and H. Chang (1980). Environmental assessment of stillage control. In *Proceedings, 2nd U.S. Department of Energy Environmental Control Symposium.* CONF-800334/2.

Bashford, L. L., and N. W. Sullivan (1981). *Conversion of a gasoline spark ignition engine to alcohol fuel.* ASAE paper 81-3583. St. Joseph, Mich.: American Society of Agricultural Engineers.

Bassett, K., and T. Chisholm (1980). *Modification of an Oliver 1550 gasoline tractor for operation with ethanol.* ASAE paper NCR80-203. St. Joseph, Mich.: American Society of Agricultural Engineers.

Bolt, J. A. (1980). A survey of alcohol as a motor fuel. In *Alcohols as motor fuels*, 21–30. Progress in Technology Series, No. 19. Warrendale, Pa.: Society of Automotive Engineers.

Borges, J. M. M. (1983). Brazil's alcohol: What's the hitch? *Sugar Journal* 45 (11): 23.

Boruff, P. A., C. F. Goering, A. W. Schwab, and E. H. Pryde (1980). *Engine evaluation of diesel fuel-aqueous ethanol microemulsions.* ASAE paper 80-1523. St. Joseph, Mich.: American Society of Agricultural Engineers.

Bothast, R. J., and R. W. Detroy (1981). What is alcohol? How is it made? In Proceedings, Alcohol and Vegetable Oils as Alternate Fuels Workshop. West Lafayette, Ind.: Purdue University.

Braun, D. E., and K. Q. Stephenson (1982). *SI engine optimized for aqueous ethanol fuel.* ASAE paper 82-3100. St. Joseph, Mich.: American Society of Agriculture Engineers.

Bryan, W. L., G. E. Monroe, R. L. Nichols, and G. J. Gascho (1981a). *Evaluation of sweet sorghum for fuel alcohol.* ASAE paper 81-3571. St. Joseph, Mich.: American Society of Agricultural Engineers.

Bryan, W. L., G. L. Newton, and J. S. Johnson (1981b). *Mechanical dewatering and storage of grain distillery waste.* ASAE paper 81-6005. St. Joseph, Mich.: American Society of Agricultural Engineers.

Butler, R. H., and L. Crombie (1980). Abstract. In *Proceedings of the Bio-Energy '80 World Congress and Exposition*, 427–29. Washington, D.C.: The Bio-Energy Council.

Cruz, J. M., C. A. Rotz, and D. H. Watson (1982). Dual-fueling turbocharged diesel with ethanol. *Transactions of the ASAE* 25 (5): 1163–68.

Daeschel, M. A., J. O. Mundt, and I. E. McCarty (1981). Microbial changes in sweet sorghum (sorghum bicolor) juices. *Applied and Environmental Microbiology* 42 (2): 381–82.

Detroy, R. W., R. L. Cunningham, and A. I. Herman (1982). Fermentation of wheat straw xylans to ethanol by *Pachysolen tannophilus*. *Biotech. and Bioeng. Symp*. 12:81–90.

Downs, H. W., B. L. Clary, and J. M. O'Neal (1982). *Enzyme function and control during conversion of grain starches to fermentable sugars*. ASAE paper 82-3597. St. Joseph, Mich.: American Society of Agricultural Engineers.

Durso, D. F. (1983). Cellulose derivatives: Arranging for the future. In E. J. Soltes, ed., *Wood and agricultural residues: Research on use for feed, fuels and chemicals*. New York: Academic Press.

Dyer, G. H. (1983). The outlook for fuel alcohol. In *Alcohol as an octane enhancer and profit extender*, Washington, D.C.: Alcohol Week and the Renewable Fuels Association.

Eakin, D. E., J. M. Donovan, G. R. Cysewski, S. E. Petty, and J. V. Maxham (1981). Preliminary evaluation of alternative ethanol/water separation processes. Pacific Northwest Laboratory, Battelle Memorial Institute, Report No. PNL-3823. Prepared for the U.S. Department of Energy under Contract DE-AC06-76RLO 1830.

Egg, R. P., C. G. Coble, and H. P. O'Neal (1982). *Alcohol production from fermentation of sweet potatoes*. ASAE Paper No.82-3599. St. Joseph, Mich.: American Society of Agricultural Engineers.

Eiland, B. R., J. E. Clayton, and W. L. Bryan (1982). *Losses of fermentable sugars in sweet sorghum during storage*. ASAE paper 82-3109. St. Joseph, Mich.: American Society of Agricultural Engineers.

Gasohol U.S.A. (1981). Applications of molecular sieve technology. *Gasohol U.S.A.* 3 (6): 10–12.

Gird, J. W. (1980). On-farm production and utilization of ethanol fuel. In *Agricultural Engineering Facts*. College Park, Md.: Cooperative Extension Service, Agricultural Engineering Department, University of Maryland.

Goldstein, I. S. (1974). The potential for converting wood into plastics and polymers and into chemicals for the production of these materials. NSF-RANN Report. Washington, D.C.: National Science Foundation.

——— (1977). The place of cellulose under energy scarcity. In J. C. Arthur, Jr., ed., *Cellulose chemistry and technology*. ACS Symposium Series 48. Washington, D.C.: American Chemical Society.

——— (1978). Chemicals from wood: Outlook for the future. Paper presented at Eighth World Forestry Congress, Jakarta, Indonesia.

——— (1981a). Composition of biomass. In I. S. Goldstein, ed., *Organic chemicals from biomass*, 9–18. Boca Raton, Fla.: CRC Press.

——— (1981b). Chemicals from cellulose. In I. S. Goldstein, ed., *Organic chemicals from biomass*, pp. 101–24. Boca Raton, Fla.: CRC Press.

——— (1983). Acid processes for cellulose hydrolysis and their mechanisms. In E. J. Soltes, ed., *Wood and agricultural residues: Research on use for feed, fuels and chemicals*, 315–28. New York: Academic Press.

Hall, J. A., J. F. Saeman, and J. F. Harris (1956). Wood saccharification: A summary statement. *Unasylva* 10:7–32.

Han, Y. W., E. A. Catalano, and A. Ceigler (1983). Treatments to improve the digestibility of crop residues. In E. J. Soltes, ed., *Wood and agricultural residues: Research on use for feed, fuels and chemicals*, 217–38. New York: Academic Press.

Harris, W. H., and R. R. Davison (1980), Chemical Engineering Department, Texas A&M University, College Station, Tex. Private communication.

Hengstebeck, R. J. (1961). *Distillation principles and design procedures*. New York: Reinhold.

Hiler, E. A., ed. (1982). *Ethanol production in small-to-medium size facilities*. Final Report, Project No. 80-B-1-1, to Texas Energy and Natural Resources Advisory Council, Austin, Tex. College Station, Tex.: Agricultural Engineering Department, Texas A&M University.

Holland, C. D. (1975). *Fundamentals and modeling of separation processes*. Englewood Cliffs, N.J.: Prentice-Hall.

Huber, J. T., A. Hargreaves, C. O. L. E. Johnson, and A. Shanan (1983). Upgrading residues and by-products for ruminants. In E. J. Soltes, ed., *Wood and agricultural residues: Research on use for feed, fuels and chemicals*, 203–16. New York: Academic Press.

Jeffries, T. W. (1982). A comparison of *Candida tropicalis* and *Pachysolen tannophilus* for the conversion of xylose to ethanol and other products. *Biotech. Bioeng. Symp.* 12:103–10.

Klausmeier, W. H. (1983). Configurations for a forest refinery. In E. J. Soltes, ed., *Wood and agricultural residues: Research on use for feed, fuels and chemicals*, 567–90. New York: Academic Press.

Ladisch, M. R., and K. Dyck (1979). Dehydration of ethanol: New approach gives positive energy balance. *Science* 205:898.

Lastick, S. M., D. Spindler, and K. Grohmann (1983). Fermentation of soluble cello-oligodextrins by yeasts. In E. J. Soltes, ed., *Wood and agricultural residues: Research on use for feed, fuels, and chemicals*, 239–52. New York: Academic Press.

Kempton, T. J., J. V. Nolan, R. A. Leng (1977). Principles for use of non-protein nitrogen and bypass proteins in the diets of ruminants. *World Anim. Rev.* 22:2.

Lee, T. S., D. Omstead, N. H. Lu, and H. P. Gregor (1981). Membrane separation in alcohol production. *Annals, New York Academy of Sciences* 369 (June 29).

Lipinsky, E. S. (1983). Disruption and fractionation of lignocellulose. In E. J. Soltes, ed., *Wood and agricultural residues: Research on use for feed, fuels and chemicals*, 489–502. New York: Academic Press.

Lockeretz, W. (1980). *Grain production for alcohol fuels*. NAFC/80-19. Washington, D.C.: National Alcohol Fuels Commission.

Marchessault, R. H., and S. Malhotra (1983). The wood explosion process: Characterization and uses of lignin and cellulose. In: E. J. Soltes, ed., *Wood and agricultural residues: Research on use for feed, fuels and chemicals*, 401–14. New York: Academic Press.

Mathewson, S. W. (1980). *Manual for the home and farm production of alcohol fuel*. Los Banos, Calif.: J. A. Diaz.

Meade, G. P., and J. C. P. Chen (1977). *Cane sugar handbook*. 10th ed. New York: Wiley.

Most, W. J., and J. P. Longwell (1980). Single-cylinder engine evaluation of methanol-improve energy economy and reduced NO_x. In *Alcohols as motor fuels*, 82–93. Progress in Technology Series, No. 19. Warrendale, Pa.: Society of Automotive Engineers.

Muzzy, J. D., R. S. Roberts, C. Fieber, S. Fass, and T. Mann (1983). Pretreatment of hardwood by continuous steam hydrolysis. In E. J. Soltes, ed., *Wood and agricultural residues: Research on use for feed, fuels and chemicals*, 351–68. New York: Academic Press.

National Research Council (1981). *Feeding value of ethanol production byproducts*. NRC Task Force of Feeding Value of Ethanol Production By-Products. Washington, D.C.: National Academy of Sciences.

Nuese, G. A., and D. R. Hunt (1982). *Sweet sorghum juice harvesting*. ASAE paper 82-1038. St. Joseph, Mich.: American Society of Agricultural Engineers.

Pollock, J. R. A. (1962). The nature of the malting process. In A. H. Cook, ed., *Barley and malt: Biology, biochemistry, technology*, 303–309. Brewing Industry Research Foundation, Redhill, England. New York and London: Academic Press.

Potter, F. L. (1983). U.S. fuel ethanol supplies, demand and price. In *Alcohol as an octane enhancer and profit extender*, Washington, D.C.: Alcohol Week and the Renewable Fuels Association.

Pringle, H. C., and M. M. Mayeux (1982). *Extending diesel fuel by carbureting aqueous ethanol*. ASAE paper 82-3101. St. Joseph, Mich.: American Society of Agricultural Engineers.

Reidenbach, V. G., and C. G. Coble. (1982). *Sugarcane or sweet sorghum processing techniques*. ASAE paper 82-3562. St. Joseph, Mich.: American Society of Agricultural Engineers.

Rein, B. K., R. L. Ogden, C. E. Walker, and D. D. Schulte (1982). *Effects of cooling on sweet sorghum juice fermentation*. ASAE paper 82-3602. St. Joseph, Mich.: American Society of Agricultural Engineers.

Samuels, G. (1980). Rum distillery wastes: Potential agricultural and industrial uses in Puerto Rico. *Sugar Journal* 43 (4): 9–12.

Sarkanen, K. V. (1980). Acid-catalyzed delignification of lignocellulosics in organic solvents. In K. V. Sarkanen and D. A. Tillman, eds., *Progress in biomass conversion*, 2:127–44. New York: Academic Press.

Schelling, G. T., J. M. Sweeten, L. M. Schake, and C. E. Coppock (1983). Digestibility of grain sorghum stillage and cottonseed hulls. In *Beef Cattle Research in Texas—1983*. College Station, Tex.: Texas Agricultural Experiment Station, Texas A&M University System.

Schrock, M. D. (1980). Using alcohol in engines. In *Proceedings, Forum on Alcohol Fuels*. Manhattan, Kan.: Kansas State University.

Seeley, D. B. (1976). Cellulose saccharification for fermentation industry applications. *Biotech. Bioeng. Symp.* 6:285–92.

Seely, R. J., and D. D. Spindler (1981). Anaerobic digestion of thin stillage from a farm-scale alcohol plant. Paper presented at fifth annual Symposium on Energy from Biomass and Wastes, Institute of Gas Technology, Lake Buena Vista, Fla.

Selke, S., M. C. Hawley, and D. T. A. Lamport (1983). Reaction rates for liquid phase HF saccharification of wood. In E. J. Soltes, ed., *Wood and agricultural residues: Research on use for feed, fuels and chemicals*, 329–50. New York: Academic Press.

Shelton, D. P., and A. R. Rider (1979a). Ethanol production equipment and processes. In *Ethanol production and utilization for fuel*, 23–28. Lincoln, Neb.: Cooperative Extension Service, University of Nebraska-Lincoln.

——— (1979b). Distillation principles. In *Ethanol production and utilization for fuel*, 20–23. Lincoln, Neb.: Cooperative Extension Service, University of Nebraska-Lincoln.

Shropshire, G. J., and C. E. Goering (1982). Ethanol injection into a diesel engine. *Transactions of the ASAE* 25 (3): 570–75.

Simmons, D. (1983). Methanol supply, demand and price. In *Alcohol as an octane enhancer and profit extender*. Washington, D.C.: Alcohol Week and the Renewable Fuels Association.

Soltes, E. J. (1980). Alternate feedstocks for ethanol production. In *Proceedings, Alcohol Fuels Symposium*. College Station, Tex.: Center for Energy and Mineral Resources, Texas A&M University.

——— (1983). Cellulose: Elusive component of the plant cell wall. In W. A. Cote, Jr., ed., *Biomass utilization*, 271–98. New York: Plenum Press.

Sorel, E., S. M. Woodward, and J. Preston (1907). *Carbureting and combustion in alcohol engines*. London: Wiley.

Strong, R. M. (1909). *Commercial deductions from comparisons of gasoline and alcohol tests on internal combustion engines*. Bulletin 392. Washington, D.C.: Department of the Interior.

Sullivan, N. W. (1981). *Pre-turbocharger alcohol fumigation in a diesel tractor*. ASAE paper 81-3581. St. Joseph, Mich.: American Society of Agricultural Engineers.

Sweeten, J. M., C. G. Coble, and R. P. Egg (1982a). The use of grain sorghum in integrated food and fuel production systems. Paper presented at the fifth Miami International Conference on Alternate Energy Sources, Clean Energy Research Institute, University of Miami, Coral Gables, Florida.

Sweeten, J. M., J. T. Lawhon, G. T. Schelling, T. R. Gillespie, H. P. O'Neal, V. G. Reidenbach, C. G. Coble, E. A. Hiler, and R. P. Egg (1982b). Nutrient recovery and pollutant control from ethanol stillage. In *Proceedings, 6th Annual Conference on Energy from Biomass and Wastes*. Lake Buena Vista, Fla.: Institute of Gas Technology.

Sweeten, J. M., J. T. Lawhon, G. T. Schelling, and R. S. Etheredge (1983a). Grain sorghum stillage processing: Nutrient recovery and pollution control. *Transactions of the ASAE* 26 (4): 1158–65, 1170.

Sweeten, J. M., J. T. Lawhon, G. T. Schelling, T. R. Gillespie, and C. G. Coble (1983b). Removal and utilization of ethanol stillage constituents. *Energy in Agriculture* 1:331–45.

Thompson, N. S. (1983). Hemicellulose as a biomass resource. In E. J. Soltes, ed., *Wood and agricultural residues: Research on use for feed, fuels and chemicals*, 101–20. New York: Academic Press.

Turbak, A. F. (1983). Newer cellulose solvent systems. In E. J. Soltes, ed., *Wood and agricultural residues: Research on use for feed, fuels and chemicals*, 87–100. New York: Academic Press.

Walter, J. B., and K. R. Kaufman (1980). *Dual fueling a farm tractor with ethanol*. ASAE paper 80-1050. St. Joseph, Mich.: American Society of Agricultural Engineers.

Ward, G. M., and J. K. Matsushima (1980). *Feeding value of distiller's dried grains (DDG) in beef feedlot rations*. Final report, SERI/TR-98340-1. Golden, Colo.: Solar Energy Research Institute.

Wu, Y. V., K. R. Sexson, and J. S. Wall (1980). Protein-rich residue from grain alcohol distillation: Fractionation and centrifugation. Paper presented at the 65th annual meeting, American Association of Cereal Chemists. Chicago, Ill.

Yang, R. D., D. A. Grow, and W. E. Goldstein (1981). *Pilot plant studies of ethanol production from whole ground corn, corn flour and starch*. Elkhart, Ind.: Miles Laboratories, Inc., Chemical Engineering Research, Biotechnology Group.

CHAPTER 5

Plant Oil Extraction and Utilization

CADY R. ENGLER, LAWRENCE A. JOHNSON, WAYNE A. LEPORI, and C. MICHAEL YARBROUGH

Plant oil was one of the early fuels tested in the compression ignition engine invented by Dr. Rudolph Diesel in the late 1800s. One of Diesel's engines on display at the Paris Exposition of 1900 was fueled with peanut oil (Grinaker, 1981). From that time until the early 1950s, numerous studies were made to evaluate plant oils as fuels for compression ignition engines. Varying degrees of success were achieved when substituting plant oils either completely or partially for diesel fuel (Wiebe and Nowakowska, 1949). The parameters evaluated most frequently were thermal efficiency and power output relative to diesel fuel, since those could be determined in short-term tests. Some reports indicated that performance with plant oils was equivalent to or better than diesel; other reports showed the opposite. In several cases, severe problems with carbon deposits, corrosion, and wear were reported. No general trends regarding the suitability of plant oils as alternative diesel fuels emerged from those studies.

A major reason for unclear or even contradictory conclusions from early studies is that very few data were recorded regarding the chemical and physical characteristics of the plant oils tested; in most cases only the type of oil (peanut, cottonseed, palm, or the like) was reported, not the extent of refining. In addition, there was greater variability in the construction of test engines than has been the case more recently; instrumentation was less accurate; and operating conditions such as ambient temperature and barometric pressure, which are now known to affect performance, were not reported.

Because of the oil embargo and rapidly escalating prices of diesel fuel in the early 1970s, interest in plant oils as alternative diesel fuels reemerged. Since many problems associated with the use of plant oil fuels already were known, recent studies have been aimed at identifying the characteristics of those fuels that are harmful to engines and developing techniques to reduce problems, by either processing the plant oils or altering the engine operating conditions. The most common vegetable oils have been subjected to standard ASTM tests for diesel fuels along with extensive chemical characterization (Goering et al., 1982a), and many recent studies have been conducted to determine performance characteristics of well-defined plant oil fuels in modern types of engines (Adams et al., 1983; Bruwer et al., 1980; Engler et al., 1983a; Peterson, Auld, and Korus, 1983; Strayer, Blake, and Craig, 1983; Ziejewski and Kaufman, 1982).

The major problems in using unmodified plant oils as alternative diesel fuels have been summarized as follows (Lipinsky et al., 1982):

(1) clogging of fuel lines and filters with fines, phosphatide gums, waxes, or high-melting-point unsaturated fats;

(2) polymerization or partial oxidation of oil during storage, which may increase viscosity or cause additional clogging problems;

(3) thermal or free radical polymerization in the combustion chamber, causing varnish formation and other deposits on cylinder walls and pistons;

(4) poor ignition and combustion characteristics caused by improper atomization;

(5) increased coke formation;

(6) polymerization of fuel blowby, which causes thickening and/or increased solids contamination of the lubricating oil.

These problems in general are caused by the higher viscosity of plant oils, by the presence of highly reactive unsaturated fatty acids as components of the triglycerides, and by the presence of minor components such as phosphatides. Problems caused by high viscosity may be lessened by preheating the oil or by interesterification with monohydroxy alcohols.

Plant and animal oils generally considered for substitute diesel fuels are composed primarily of triglycerides, which are branched molecules having approximately three times the molecular weight of typical diesel fuel components. These oils are divided into groups according to chemical composition, as shown in Table 5.1 (Sonntag, 1979). The iodine value (IV) is a measure of the amount of unsaturation (double bonds) in an oil, and the saponification number is inversely proportional to the average molecular weight of an oil. Oils produced extensively in the U.S. belong to the following groups:

TABLE 5.1 Typical Characteristics of Major Fats and Oils

				Fatty Acid Composition (%)[c]		
Group	Fat or Oil	IV[a]	SN[b]	12:0	14:0	16:0
Milk fat	Butterfat	33	225	3.1	11.7	26.2
Lauric acid	Coconut	10	256	48.5	17.6	8.4
Vegetable butter	Cocoa butter	36	200		0.1	25.8
Animal fat	Beef tallow	41	196	0.1	3.3	25.5
Oleic-linoleic acid	Cottonseed	110	194		0.9	24.7
	Peanut	93	192		0.1	11.6
	Sunflower	130	191	0.5	0.2	6.8
	Safflower	145	192		0.1	6.5
Erucic acid	Rapeseed	103	176		0.1	2.8
Linolenic acid	Soybean	131	192		0.1	11.0
	Linseed	185	192			7.0
Conjugated acid	Tung	167	192			3.1
Marine animal	Menhaden	157	195		7.2	17.0
Hydroxy acid	Castor	86	181			1.6

(a) Iodine value
(b) Saponification number
(c) $x{:}y$ indicates x carbon atoms in the fatty acid chain with y double bonds

(1) oleic-linoleic acid oils (cottonseed, sunflower, and peanut); (2) linolenic acid oils (soybean); (3) animal fats (beef tallow and greases). Plant oils not currently produced extensively in the U.S. (for example, coconut, palm, and castor) could become important alternative diesel fuels in other countries.

Typical fuel properties of plant oils and diesel fuel are shown in Table 5.2. Cloud points (temperatures at which crystallization begins)

Fatty Acid Composition (%)[c]							
16:1	18:0	18:1	18:2	18:3	20:0	20:1	Unique Fatty Acids
1.9	12.5	28.2	2.9	0.5		0.2	C_4, C_6, C_8, C_{10} acids, 9.2%
	2.5	6.5	1.5		0.1		C_8, C_{10} acids, 14.4%
0.3	34.5	35.3	2.9		1.1		
3.4	21.6	38.7	2.2	0.6	0.1		
0.7	2.3	17.6	53.3	0.3	0.2	1.3	
0.2	3.1	46.5	31.4		1.5	1.4	
0.1	4.7	18.6	68.2	0.5	0.4		
	2.4	13.1	77.7		0.2		
0.2	1.3	23.8	14.6	7.3	0.7	12.1	Erucic acid (22:1), 34.8%
0.1	4.0	23.4	53.2	7.8	0.3		
	4.0	15.0	18.0	55.0			
	2.1	11.2	14.6	69.0			18:3 has conjugated double bonds
7.9	4.0	13.4	1.1	0.9		0.9	highly unsaturated, long-chain acids (20:5, 10.2%; 22:6, 12.8%)
	1.8	5.7	6.7	0.3	0.5		12-hydroxy-9-octadecenoic acid, 83.6%

for plant oils and plant oil esters are somewhat higher than for diesel fuel, which could cause problems of sedimentation and filter plugging during cold periods. On the other hand, flash points for plant oil fuels are significantly higher than for diesel fuel, making them less hazardous to handle. Viscosities of plant oils are high, as mentioned previously, but plant oil esters have viscosities similar to that of diesel fuel. Heating values for plant oil fuels are approximately 10 percent lower than for diesel fuel, which causes higher specific fuel consumption. Cetane numbers, when measured for plant oil fuels using standard ASTM procedures, are in a range of 30 to 40, about 10 units below the range for normal hydrocarbon fuels; however, results from other methods of rating combustion characteristics indicate that the cetane method underrates the combustion properties of plant oils (Bacon et al., 1981; Fort et al., 1982; Ryan, Callahan, and Dodge, 1982).

This chapter examines oilseed processing technology (extraction and refining) as it applies to fuel preparation, engine test results, and chemical modification to produce better fuels.

Extraction Technology

Current practice in the oilseed milling (extraction) industry is to use either screw pressing (expelling) or solvent extraction to separate the oil from oil-bearing seeds. The objectives of both processes are to obtain a high-quality oil as free from impurities as possible, to obtain the oil in high yield, and to produce a high-quality cake or meal residue suitable for animal feed or for further processing to protein isolates and concentrates (Norris, 1982a). To meet these objectives, careful preparation of the seed for expression or extraction is required. An extensive review of commercial-scale oilseed extraction is given by Norris (1982a). For small-scale extraction of oil for fuel, many of the steps described below can be omitted, but at the cost of significantly reduced oil yields.

Seed Preparation

Seed preparation steps will vary, depending on the type of seed, as shown in the comparison of soybeans and cottonseed given in Figure 5.1. Extraneous materials such as sticks and other plant trash, dirt, and sand must be removed to reduce product contamination and damage to subsequent equipment. This is generally accomplished by screening and aspiration. Magnets are used to remove iron.

The hulls of oilseeds are low in oil content. If not removed prior to the extraction process, hulls absorb and retain oil in the meal, reduce the capacity of extraction equipment, and may contribute such undesirable contaminants as waxes. In addition, hulls passing through the extraction step dilute the protein content of the meal. Cottonseed and sunflower seed generally are decorticated with bar or disc hullers.

TABLE 5.2 Typical Fuel Properties of Plant Oils and Diesel Fuel

Fuel	Specific Gravity	Cloud Point (°C)	Flash Point (°C)	Viscosity at 40°C (mPa·s)	Heating Value (MJ/kg)	Ash (%w)	Phos-phorus (ppm)	Sulfur (%)	Cetane No.
No. 2 diesel	0.83	−13	56	2	45.0	0.02	—	<0.5	>40
Cottonseed oil									
Crude	0.92	—[a]	233	31	40.3	0.01	300	<0.01	—[b]
Degummed	0.92	—[a]	224	31	40.3	0.01	50	<0.01	—[b]
Alkali refined	0.92	2	242	31	40.3	0.002	0	<0.01	—[b]
Sunflower oil									
Degummed	0.92	9	257	31	39.3	0.01	190	<0.01	—[b]
Soybean oil									
Ethyl esters	0.88	8	182	4	40.1	<0.01	0	<0.01	—[b]

(a) Not measurable because of pigments in oil.
(b) Cetane number does not appear to be meaningful for plant oil fuels.

In both types, hulls are split as seeds are caught between a rotating cutting edge and an opposing stationary edge. Hulls are then separated from kernels by various combinations of vibrating screens and aspirators, using a system such as that shown in Figure 5.2.

Soybeans need not be decorticated as the hulls are a very small fraction of the seed and are relatively nonabsorbent. However, soybean dehulling is common practice today; it is accomplished by first cracking the beans, using cracking rolls, then separating the hulls by means of vibrating screens and aspiration.

The reduction of seeds to small particles facilitates the extraction of oil by either mechanical expression or solvents. To prepare seeds for mechanical expression, a series of five crushing rolls placed one above another is used. Seed kernels pass four times between rolls, resulting in considerable breakage and the formation of flakes. To prepare flakes for solvent extraction, a pair of rolls mounted horizontally is used. Clearance between the rolls can be adjusted to produce flakes of uniform thickness, and the less intense action provides large, coherent flakes with fewer fines.

Oilseeds are heat treated, or cooked, prior to extraction to coagulate proteins and make the cell wall structures of fat-containing cells permeable to oil. Heating ruptures the cell walls by thermal stress and facilitates the flow of oil from the solids by reducing viscosity. Cooking also may inactivate the toxic components present in some oilseeds, reduce the solubility of phosphatides, and reduce the affinity of solid surfaces for oil. In preparing seed for solvent extraction, the cooking is generally done prior to flaking, to aid in the formation of thin, coherent flakes.

Screw Pressing

Screw pressing is the simplest means of extracting oil from oilseeds and has been given serious consideration for small-scale production of oil as an alternative diesel fuel. In screw presses, cooked oilseed flakes are subjected to increasing pressure as they are conveyed through a barrel cage using a tapered screw (Figure 5.3). The barrel cage is usually made up of rectangular bars spaced up to 0.25 mm apart to allow drainage of oil from the barrel. Bar spacing may vary along the length of the barrel and for the type of seed being pressed. Screw design also may vary. Cake is discharged from the end of the

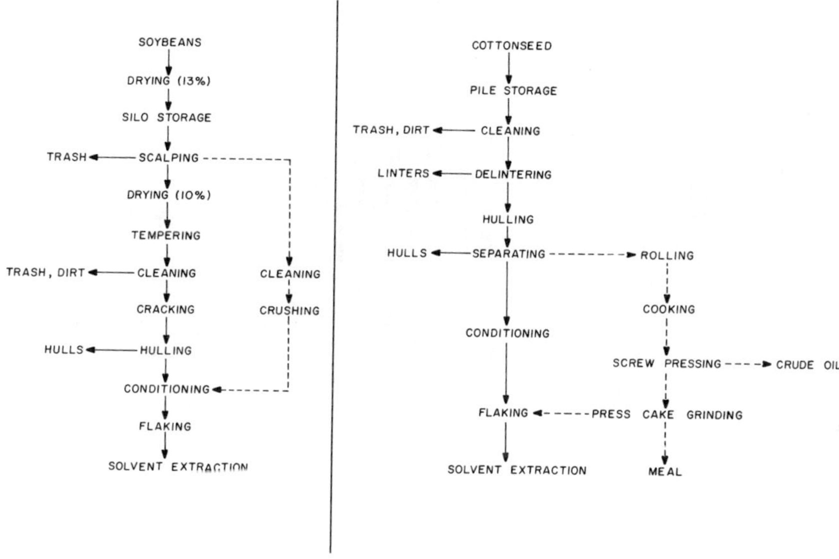

FIG. 5.1. Process flow diagrams for soybean and cottonseed oil extractions.

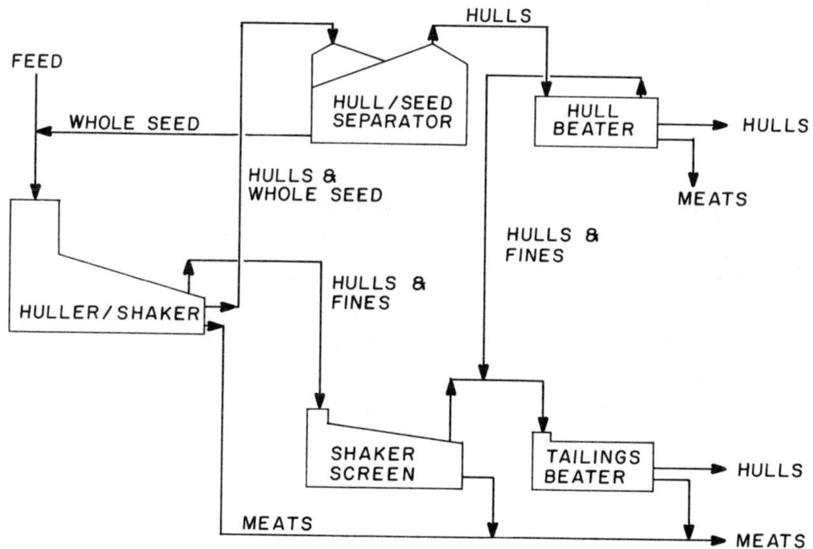

FIG. 5.2. Typical system for separation of hulls from oilseed kernels.

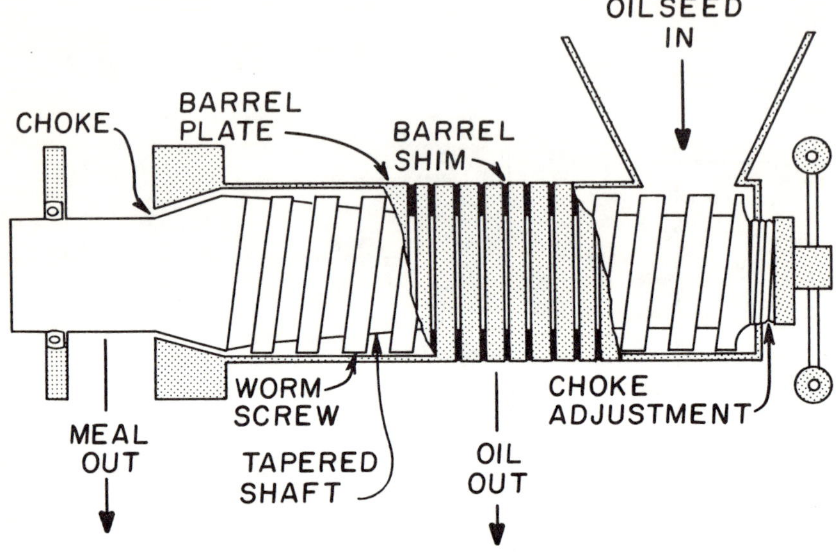

FIG. 5.3. Typical screw press for oilseed extraction.

barrel past a choke device that regulates the pressure generated in the barrel.

Under optimum operating conditions, screw presses yield cake with a residual oil content of only 3 to 4 percent; however, small-scale presses, which are much less efficient, generally yield a cake with 12 percent or greater oil content. The lower efficiency of small presses is due in part to nonoptimal seed preparation and in part to the fact that those presses are not designed to operate at the high pressures required for hard-pressing of oilseed. Residual oil content of the cake is a function of seed preparation and press operation; it does not depend on the amount of oil initially present in the seed.

High-oil-content seed may be processed by a combination of screw pressing and solvent extraction: seed is pressed to a residual oil content of approximately 12 percent, and the press cake is then solvent extracted as below.

Solvent Extraction

While screw pressing gives a high recovery of oil from kernels initially having a high oil content, as much as 20 percent of the total oil may be left in the cake of low-oil-content seeds such as soybeans. Solvent extraction has been developed to increase the recovery of oil. Typically, meal from a solvent extraction process has a residual oil content less than 1 percent.

Although a number of different solvent extraction systems have been developed, commercial plants in the U.S. use continuous countercurrent percolation extractors following the process flow scheme shown in Figure 5.4. In these systems, flakes entering the extractor are sprayed with a relatively concentrated miscella (oil-solvent mixture), which percolates by gravity through the bed of flakes. The bed of flakes is contacted by successively less concentrated miscellas until near the exit it is contacted with fresh solvent. Operating temperatures generally are 2 to 5°C below the solvent boiling point. The average capacity for an extractor is 750 tons per day, although some extractors can process up to 3,000 tons per day.

The most common solvent for oilseed extraction is hexane (boiling range 63–69°C). Because of fire and explosion hazards, limitations on environmental emissions, and the increasing costs of hexane and other hydrocarbon solvents, there has been considerable interest in alternative solvents. Recent reviews of solvents used for oilseed extraction are given by Hron, Koltun, and Graci (1982), Johnson and Lusas (1983), and Norris (1982a). The most commonly considered alternatives to hexane are ethanol, isopropanol, acetone, and methylene chloride. Supercritical carbon dioxide also has been studied as an alternative oilseed extraction solvent (Friedrich, List, and Heakin, 1982; Mangold, 1983). Because, in addition to extracting oil, alternative solvents may extract varying amounts of carbohydrates, toxic components (such as gossypol and aflatoxin), and other components of oilseeds, some modification of oil refining and solvent recovery steps are required. An aqueous process also has been developed whereby the oil is emulsified with water and recovered by centrifugation (Rhee, Cater, and Mattil, 1972).

Oil Processing

Crude oils obtained by either screw pressing or solvent extraction contain varying amounts of impurities that may affect the oils' performance as fuel. The behavior of plant oils as fuel is considerably different from that of diesel fuel; therefore, plant oils cannot be required to

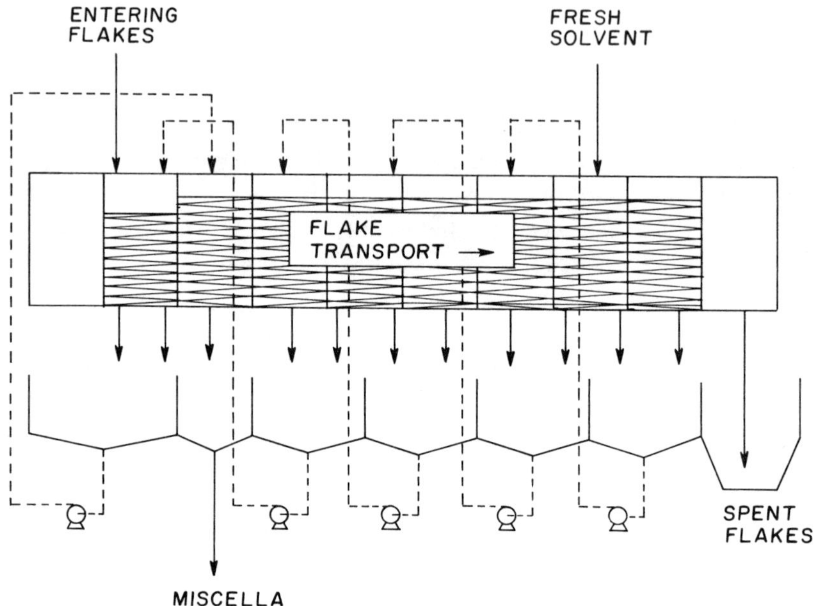

FIG. 5.4. Process flow diagram for countercurrent solvent extraction of oilseeds.

meet the same specifications. Pryde (1982) has suggested standards for plant oil fuels (Table 5.3) which include many characteristics currently used to evaluate edible oils. Further research is required to define oil characteristics and levels of impurities acceptable in fuel applications.

Crude oil may be processed to remove impurities such as particulate matter, phosphatides, waxes, pigments, and free fatty acids. Oils also may undergo chemical modification to change the degree of saturation or to convert the triglyceride structure to esters of fatty acids and monoalcohols. It is also possible to crack plant oils with the aid of shape-selective catalysts to produce essentially the equivalent of diesel fuel.

Filtration

Particulate matter must be removed from oils to prevent plugging of fuel lines and filters. Unless it is filtered, particulate matter will cause rapid buildup of carbon deposits in the engine and increased wear from abrasion. Crude oil from either screw pressing or solvent extraction contains a significant quantity of seed fragments termed *foots*. Large particles in the foots are removed by settling, and smaller particles in pressed oil are removed by filtration. Solvent-extracted oil may be processed through hydroclones or filters to remove smaller particles prior to solvent evaporation. For fuel use, the oil must be filtered to remove particulates down to approximately 4μm as a final processing step.

Degumming

Crude plant oils contain from 0.5 to 3.0 pecent phosphatides and other mucilaginous materials, which are removed by degumming. Degumming is a simple hydration proccss (Figure 5.5) that reduces phosphatide content to approximately 0.15 percent but does not significantly affect the free fatty acid content of the oil. When phosphatides are hydrated, they become insoluble in oil and can easily be removed by centrifugation. Gums in solvent-extracted oils can be hydrated with condensate formed in the final stage of the steam stripping that removes the last traces of solvent from the oil. Dry oils, as from screw pressing, are degummed by mixing them with a small amount of water (1–3 percent) and heating to 35–50°C. Although heating is not necessary for hydration to occur, it reduces oil viscosity, thus making separation of the gums easier. For small-scale degumming, hydrated gums can be allowed to settle in a tank, and the degummed oil decanted. A comprehensive review of degumming is given by Norris (1982b).

Alkali Refining

Free fatty acids are probably not detrimental to the combustion characteristics of plant oils but may cause increased corrosion problems in fuel systems and engines. Among the several methods of removing or neutralizing free fatty acids, the simplest is refining with caustic soda, as shown schematically in Figure 5.6. In this process, free fatty acids react with sodium hydroxide to form soap, which is insoluble in oil. The amount of caustic required is affected not only by the level of free fatty acids but by pigments and surface active agents present in the oil and by the methods used to process the oilseeds. A standard laboratory analysis (AOCS, 1971) is used to determine the amount of caustic needed to refine a given oil without causing exces-

TABLE 5.3 Suggested Standards for Vegetable Oil and Ester Fuel vs. Diesel Oil Specifications

Property	Test Method[a]		Suggested Standards		ASTM Specifications for No. 2 Diesel Oil (D975)
	AOCS	ASTM	Oil	Ester	
Physical					
Cetane number (min.)	—	D613	(30–40)	(30–40)	40
Cloud point (max.)	Cc6	D2500	20°C	20°C	6°C above 10th percentile minimum ambient temperature[b]
Distillation temperatures, 90% point	—	D86			
min.			(>300°C)	(>300°C)	282°C
max.					338°C
Flash point (min.)	Cc9a	D93	(~300°C)	(~300°C)	52°C
Pour point (max.)	—	D97	−5°C	−5°C	—
Specific gravity, 15/15°C	Cc10a	D1217	0.910–0.930	(0.8–0.9)	—
Turbidity	—	—	—	—	—
Viscosity, mm^2/s at 38°C	—	D445	30–50	(3–6)	1.9–4.1

Compositional					
Ash (max.)	Ca11	D482	0.05%	0.05%	0.01%
Carbon residue on 10% residuum (max.)	—	D524	—	—	0.35%
Copper strip corrosion (max.)	—	D130	—	—	No. 3
Free fatty acid (max.)	Ca5a	—	0.2%	0.2%	—
Insolubles (max.)	Ca3	D128	0.001%	0.001%	—
Iodine value	Cd1	—	80–145	80–145	—
Moisture (Karl Fischer) (max.)	Ca2e	D1744	0.2%	0.2%	—
Phosphorus (max.)	Ca12	D1091	0.02%	0.02%	—
Sulfur (max.)	—	D129	—	—	0.50%
Volatile matter (max.)	Ca2b, c, or d	—	0.3%	0.3%	—
Water and sediment (max.)	—	D1796	—	—	0.05%
Wax (max.)	—	D3117	0.02%	0.02%	—

Source: Reprinted from Pryde (1982) by courtesy of the American Society of Agricultural Engineers.

(a) AOCS: American Oil Chemists' Society; ASTM: American Society for Testing and Materials. Values in parentheses are actual values or ranges and not necessarily a suggested specification.

(b) For example, the 10th percentile minimum temperature for Northern Illinois is −9°C for November and 11°C for March, bracketing the winter period when there would be minimum tractor activity in the field.

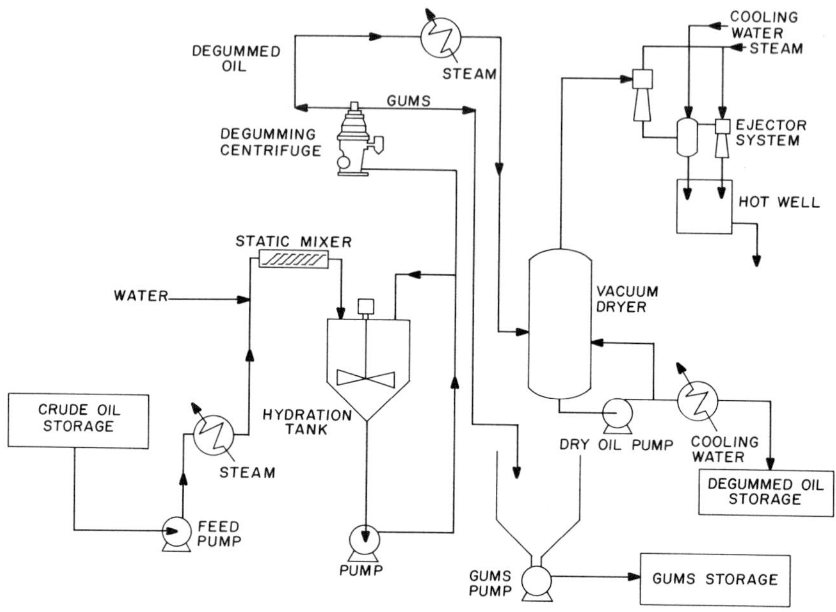

FIG. 5.5. Process flow diagram for degumming plant oils.

sive refining losses. An extensive review of alkali refining practice is given by Norris (1982b).

Alkali refining is generally carried out as a continuous process: crude oil is continuously mixed with a dilute caustic soda solution, held for a short time to allow reactions to occur, and then heated to break the emulsion that forms. Soapstock (reaction products) is separated from refined oil by continuous centrifugation; the oil is washed with hot water and centrifuged again to remove trace quantities of soap. Refined oil is finally passed through a vacuum dryer to remove traces of moisture. Since alkali refining is carried out in an aqueous environment, gums are hydrated and separated along with the soapstock, thereby eliminating the need for a separate degumming step.

For small-scale operation, alkali refining can be done in batches by mixing caustic soda solution with oil at room temperature. Agitation must be sufficiently vigorous to emulsify the mixture. When the tank contents are thoroughly emulsified, agitation is reduced and the tank contents heated to 55–60°C, causing the emulsion to break and

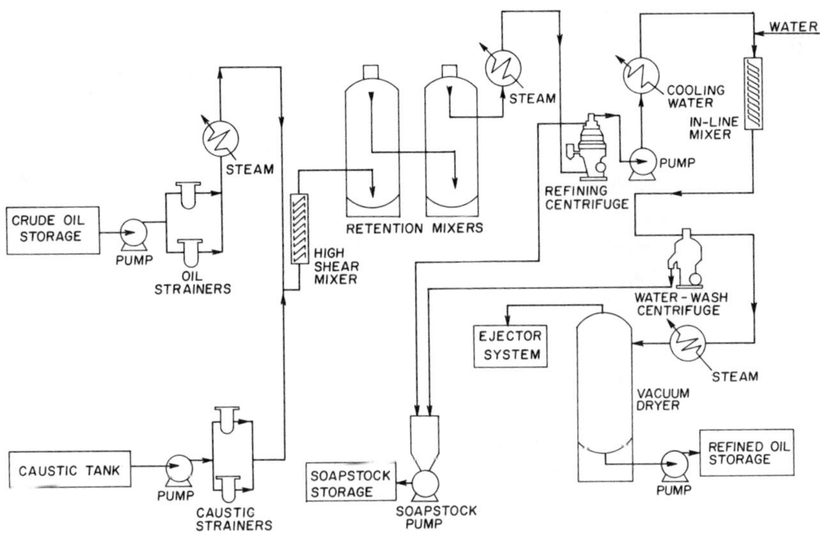

FIG. 5.6. Process flow diagram for the alkali refining of plant oils.

soapstock to coalesce. Heating and agitation are then stopped, and the soapstock is allowed to settle.

Interesterification

Plant oils have significantly higher viscosities than diesel fuel at normal injection temperatures (40–60°C), as shown in Table 5.4. This has been identified as one of the major factors causing problems in using plant oils as alternative fuels. Methyl or ethyl esters of plant oils have viscosities similar to that of diesel fuel (Table 5.4), and a report that sunflower oil ethyl esters gave good performance characteristics in engine tests (Bruwer et al., 1980) has stimulated considerable interest in interesterification of plant oils with simple monohydroxy alcohols to produce an alternative fuel.

The term *interesterification* refers to several different reactions, but for alternative fuels the only reaction of interest is the interchange of fatty acids between glycerol and low-molecular-weight monohyroxy aliphatic alcohols (methanol, ethanol, and so on). This reaction also may be called esterification, transesterification or alcoholysis. The reaction for interesterification of a plant oil with methanol is as follows:

TABLE 5.4 Viscosities of Plant Oils and Diesel Fuel at Normal Injection Temperature (40°C)

Oil	Viscosity (mPa·s)
Cottonseed	31
Peanut	36
Rapeseed	34
Soybean	30
Sunflower	31
Tallow (Beef)[(a)]	24
Cottonseed ethyl esters	4
Sunflower ethyl esters	4
Cottonseed methyl esters	4
Sunflower methyl esters	4
Diesel	3

(a) Viscosity at 45°C since solid at 40°C.

$$\begin{array}{l} CH_2O_2CR \\ | \\ CHO_2CR' \\ | \\ CH_2O_2CR'' \end{array} + 3CH_3OH \rightarrow \begin{array}{l} CH_2OH \\ | \\ CHOH \\ | \\ CH_2OH \end{array} + RCO_2CH_3 + R'CO_2CH_3 + R''CO_2CH_3$$

triglyceride + methanol → glycerol + 3 methyl esters

The conversion of plant oils to lower alcohol esters has two effects on physical properties, which lead to improved fuel characteristics. One effect is the reduction in molecular weight to approximately one-third that of the original triglyceride. The second effect is the conversion of a long, branched molecule into shorter, straight-chain molecules. Together, these effects produce significantly lower viscosities for lower alcohol esters.

Interesterification reactions between plant oils and lower alcohols can be catalyzed by either acid or alkali, but alkali catalysis is generally superior. A review of interesterification processes has been given by Sonntag (1982), and the production of lower alcohol esters for alternative fuels has been discussed by, for example, Freedman and Pryde,

1982; Hassett and Hasan, 1982; Kusy, 1982; Romano, 1982. Several alkali catalysts are effective for interesterification reactions, including potassium or sodium hydroxide, sodium methoxide, sodium ethoxide, and metallic sodium. The oil must be clean, dry, and low in free fatty acids; and the alcohol must be anhydrous. A process flow diagram for interesterification is shown in Figure 5.7. The catalyst, 0.1–0.5 percent by weight of the oil, is dissolved in alcohol, and the alcohol mixture is added to oil that has been heated to approximately 80°C. For essentially complete conversion of plant oil, the amount of alcohol added must be approximately two times the stoichiometric requirement. After thorough stirring, the mixture is allowed to stand. Glycerol begins to separate almost immediately, as it is immiscible with other components in the system. The ester layer contains the catalyst, unreacted alcohol, traces of glycerol, and soap. The unreacted alcohol may be removed by vacuum distillation for recycling, and other impurities can be washed from the ester phase with hot water.

Separation of the catalyst from lower alcohol esters to be used for fuels is imperative; otherwise, the fuel will be highly alkaline and may cause deposit formation or corrosion problems in the engine. In addition, during storage the fuel might absorb moisture from the air, which would react with the catalyst and esters to form soap.

Engine tests of plant oil esters have shown their performance to be equal to or better than that of diesel fuel without any of the engine fouling problems that have occurred with direct use of plant oils.

Although the interesterification reaction is simple enough to be carried out on a small scale, the need for strict adherence to anhydrous conditions and low free fatty acid content of the oil makes the process more appropriate for larger-scale, centralized plants. Another disadvantage of small-scale operations is the difficulty of finding a market for small quantities of glycerol, since it is normally handled as a bulk commodity. Finally, unless alcohol is stripped from the esters for recycling, which increases the complexity of the process, water washing cannot be used to remove catalyst without incurring significant alcohol losses.

Cracking

Plant oil molecules can be broken into smaller fragments by thermal or catalytic cracking. These processes give a variety of products,

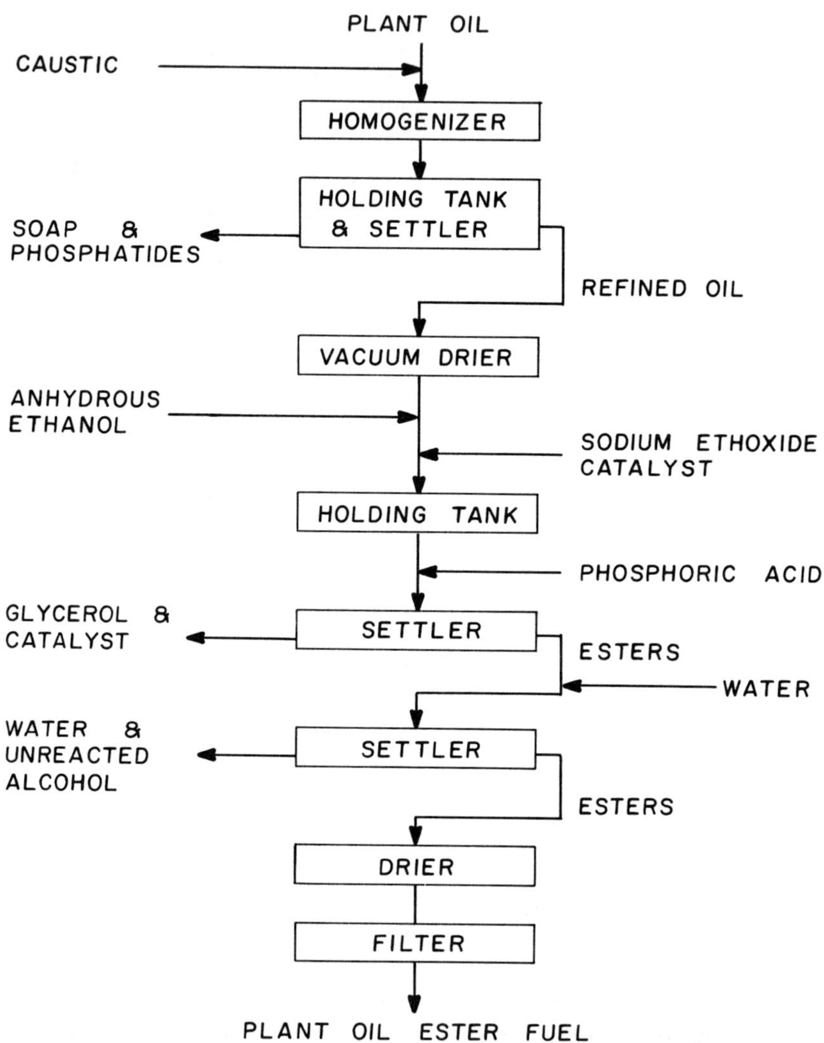

FIG. 5.7. Process flow diagram for processing crude sunflower oil into ethyl ester fuels.

including light paraffins, olefins, and aromatics; distillates in the gasoline and diesel boiling ranges; and, in some cases, tars or coke. The earliest reported cracking processes gave a wide range of products from coke to gas (Chang and Wan, 1947; Egloff and Morrell, 1932). However, recently developed processes using shape-selective zeolite catalysts give mostly light paraffins, olefins, and aromatics, with only a small amount of coke (Graille et al., 1981; Haag, Rudewald, and Weisz, 1980; Weisz, Haag, and Rudewald, 1979). Conversion of plant oils over zeolite catalysts occurs at atmospheric pressure and 450–500°C. The product composition reported for the cracking of corn oil is given in Table 5.5.

Engine Testing

In the past few years there have been numerous studies using plant oils or lower alcohol esters of plant oils as diesel fuels. These tests have identified high viscosity and unsaturation of plant oils as the primary causes of increased fouling of the engine and the lubricating oil. Short-term performance tests have been used to screen alternative fuels, and results from longer-term tests are now becoming available to identify problems related to engine durability when using plant oil fuels.

Modern diesel engines can be divided into two types: those using direct injection, and those designed with a precombustion chamber. In the first type, fuel is injected directly into the combustion chamber, as shown in Figure 5.8; in the second, fuel is injected into a small swirl chamber prior to entering the combustion chamber, as shown in Figure 5.9. Fuel entering the precombustion chamber creates a vortex which promotes atomization and dispersion of the fuel. Plant oil fuels show somewhat better performance in precombustion-chamber engines than they do in direct-injection types (Braun and Stephenson, 1982; Bruwer and Hugo, 1981; Engler et al., 1983b). Recent research with plant oil fuels has emphasized modifying the oils to work in existing engines, since it is not likely that engine manufacturers will build engines designed specifically for a limited plant oil fuel market. However, it is possible that relatively inexpensive modifications or adjustments could be made to existing engines to improve performance with plant oil fuels.

TABLE 5.5 Product Composition from Catalytic Cracking of Corn Oil

Product	Weight (%)
Methane	2.12
Ethane	2.37
Ethylene	2.53
Propane	18.49
Propylene	2.85
i-Butane	3.78
n-Butane	2.58
Butenes	0.88
C_5 nonaromatics	3.22
C_6–C_{13} nonaromatics	4.92
$C_{14}{}^+$ nonaromatics	3.50
C_6 aromatics	7.76
C_7 aromatics	22.05
C_8 aromatics	16.91
C_9 aromatics	3.53
Coke	1.51

Source: Haag, Rudewald, and Weisz (1980).

Short-Term Testing

Short-term engine tests have been used to develop performance curves for different fuels in a given test engine and to compare different engines operating on a particular fuel. The results of numerous short-term tests generally indicate that plant oils are very good alternative fuels in terms of power output and thermal efficiency. Unfortunately, sustained operation of engines with fuels containing greater than 20 percent plant oil has frequently caused increased fouling of injectors, combustion chambers, and lubricating oil. In some cases, similar problems have been noted even in short-term tests.

The plant oils most frequently considered as alternative fuels include sunflower, soybean, cottonseed, peanut, safflower, and rapeseed. From results reported to date, there appears to be little difference in short-term performance among these oils if they have been processed at least through a degumming step. Common observations from short-term tests using degummed oils as complete replacements for diesel fuel are that thermal efficiencies range from 90 to 110 percent of diesel efficiency, specific fuel consumption is 10 to 15 percent greater,

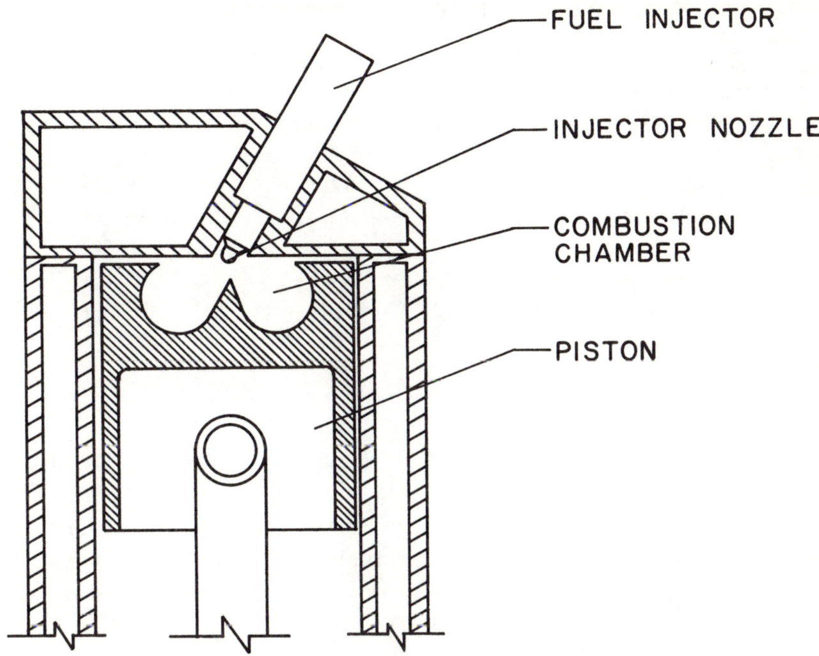

FIG. 5.8. Schematic diagram of a direct-injection diesel engine.

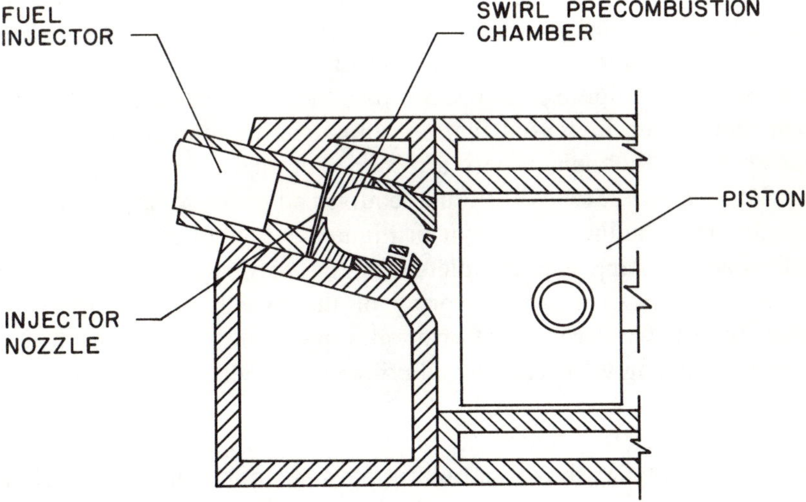

FIG. 5.9. Schematic diagram of a diesel engine with precombustion chamber.

TABLE 5.6 Heating Values of Plant Oils, Esters, and Diesel

Oil	MJ/kg	MJ/m^3
Cottonseed	40.3	37,000
Peanut	39.8	36,400
Soybean	39.2	36,000
Sunflower	39.3	36,200
Tallow (beef)	39.9	36,700
Soybean ethyl esters	40.1	35,100
Sunflower ethyl esters	41.2	36,000
Diesel	45.5	37,700

and maximum power output declines by up to 5 percent (Braun and Stephenson, 1982; Engler et al., 1983a; Humke and Barsic, 1981; Vinyard et al., 1982; Yarbrough, LePori, and Darcey, 1982). Plant oils have lower heating values than diesel fuel (Table 5.6), which accounts for most of the observed differences in specific fuel consumption and maximum power output.

Results reported by Engler et al. (1982, 1983b) for performance tests with a single-cylinder precombustion-chamber engine show that 100 percent sunflower oil fuels gave thermal efficiencies somewhat greater than diesel (Figure 5.10), while 100 percent cottonseed oil fuels gave somewhat lower efficiencies (Figure 5.11). Corresponding data for specific fuel consumption with sunflower oil (Figure 5.12) and cottonseed oil (Figure 5.13) show that the lower heating values of plant oils caused increased fuel consumption. Differences in thermal efficiency and fuel density also affect specific fuel consumption.

Performance curves for crude cottonseed oil using a precombustion-chamber engine are shown in Figures 5.11 and 5.13. Although performance testing was completed, engine performance was erratic, and heavy deposits caused scoring of the exhaust valve stem and guide. In tests with crude sunflower oil, rapid fouling of the lubricating oil caused testing to be terminated prior to completion of performance curves.

The results of cottonseed oil tests in a single-cylinder direct-injection engine are shown in Figures 5.14 and 5.15 (thermal efficiencies and specific fuel consumption, respectively). These results indi-

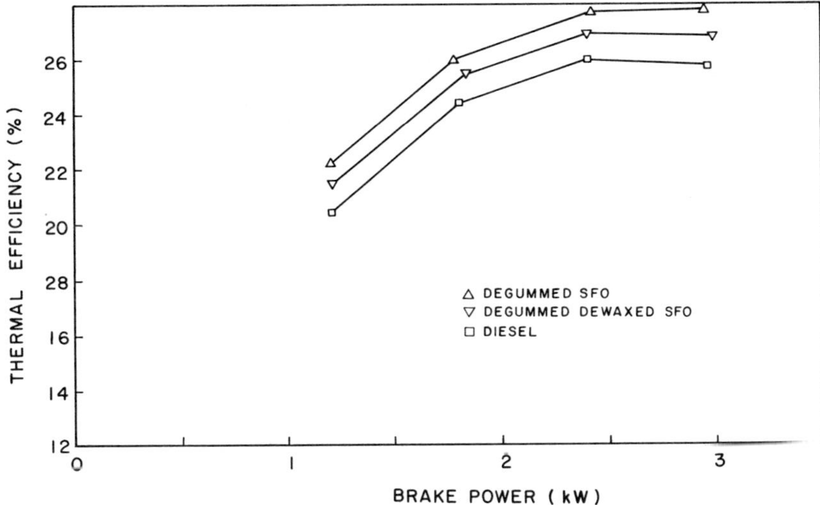

FIG. 5.10. Thermal efficiencies for sunflower oil (SFO) fuels in a single-cylinder precombustion-chamber engine.

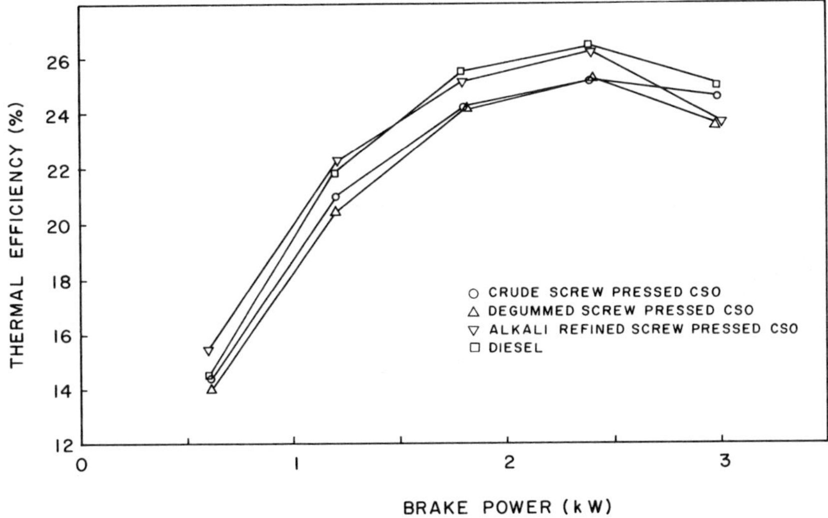

FIG. 5.11. Thermal efficiencies for cottonseed oil (CSO) fuels in a single-cylinder precombustion-chamber engine.

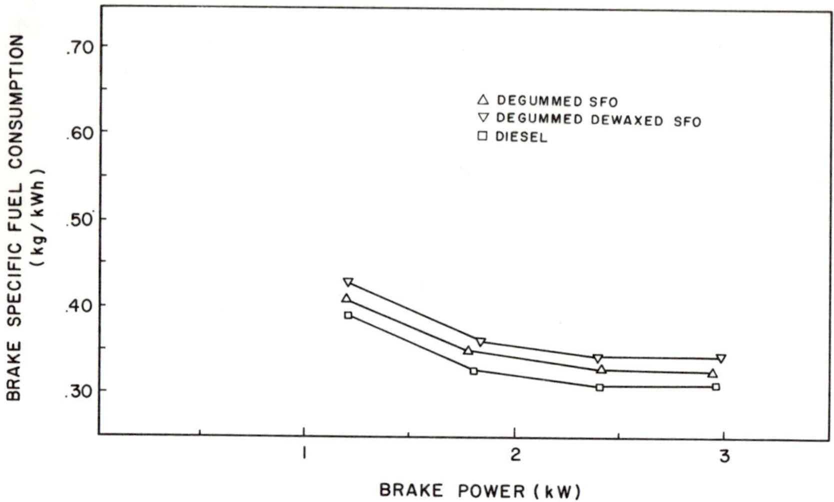

FIG. 5.12. Specific fuel consumption for sunflower oil (SFO) fuels in a single-cylinder precombustion-chamber engine.

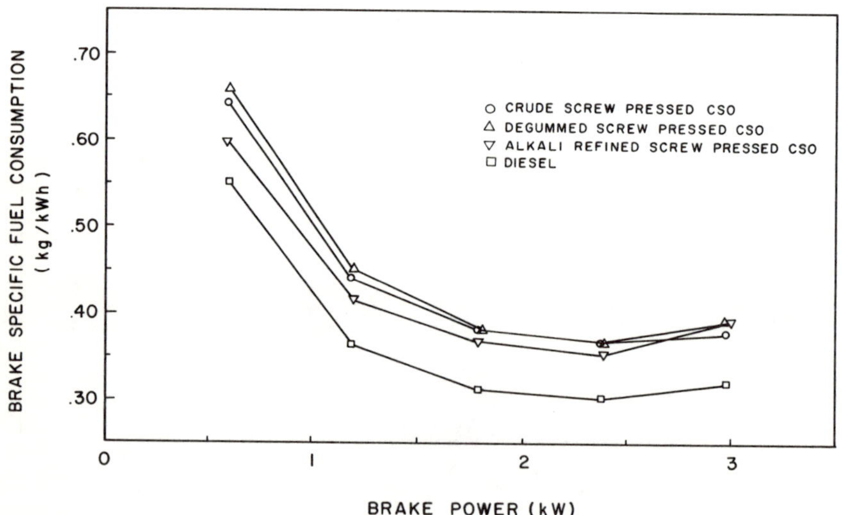

FIG. 5.13. Specific fuel consumption for cottonseed oil (CSO) fuels in a single-cylinder precombustion-chamber engine.

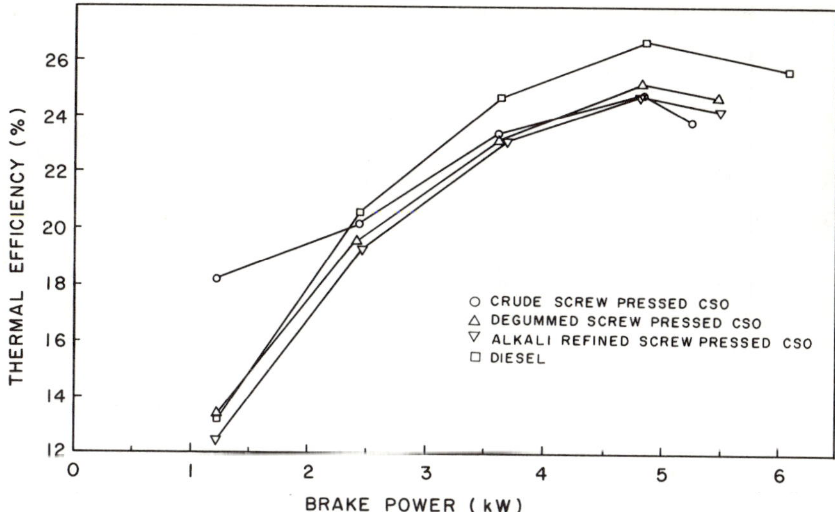

FIG. 5.14. Thermal efficiencies for cottonseed oil (CSO) fuels in a single-cylinder direct-injection engine.

cate that thermal efficiencies of cottonseed oil fuels are somewhat lower in a direct-injection engine than in a precombustion-chamber engine. Better efficiencies in the latter are apparently the result of increased mixing of fuel and air, which improves combustion.

Thermal efficiencies for blends of degummed cottonseed oil with diesel fuel are shown in Figure 5.16. These results indicate that thermal efficiencies for blends fall between efficiencies for the pure fuels roughly in proportion to the composition of the blend. Performance curves for blends of crude sunflower oil with diesel could be obtained only for the most dilute blend (25 percent sunflower oil). Blends with greater amounts of crude sunflower oil caused rapid fouling of the engine lubricating oil with subsequent termination of testing.

Inspections of injectors and the precombustion chamber after each performance test indicated that more highly processed plant oil fuels (degummed, dewaxed sunflower oil and alkali-refined cottonseed oil) caused slightly less carbon formation than less-processed oils. When blends of cottonseed oil and diesel fuel were burned, deposit formation decreased with increasing dilution of the cottonseed oil. For all test fuels, both blends and 100 percent plant oils, carbon deposits were sig-

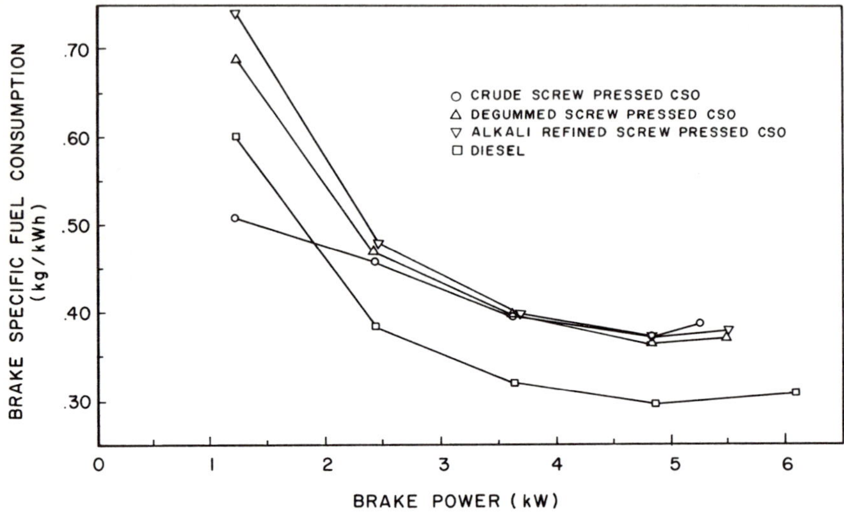

FIG. 5.15. Specific fuel consumption for cottonseed oil (CSO) fuels in a single-cylinder direct-injection engine.

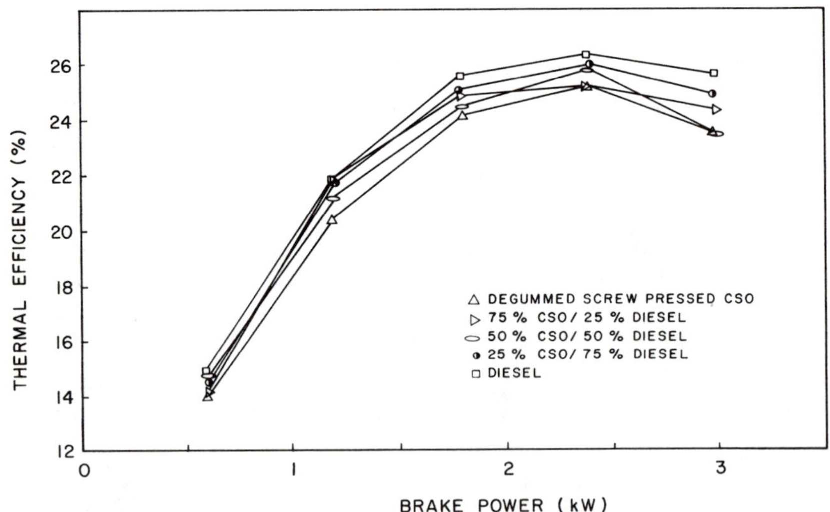

FIG. 5.16. Thermal efficiencies for blends of degummed cottonseed oil (CSO) with diesel in a single-cylinder precombustion-chamber engine.

nificantly greater than with diesel fuel, but they were easily removed and did not appear to be detrimental to engine operation.

Although 200-hour tests are recommended to evaluate engine wear, ring or valve sticking, and lubricating oil fouling (Alternate Fuels Committee, 1982), short endurance tests of 40 to 50 hours can be used for early identification of serious problems. During short endurance tests, lubricating oil samples are collected every 10 hours and analyzed for trace metals, solids content, and viscosity. Following the test, the engine is disassembled for inspection and wear measurements. Although trace metal profiles from analysis of lubricating oil samples are not conclusive, trends for various metals may be indicative of engine wear and other problems. Increasing content of metals such as iron, chromium, aluminum, lead, and copper may indicate engine wear. Depletion of antiwear additives may be indicated by reductions in phosphorus and zinc, and depletion of dispersants and detergents by reductions in calcium, barium, and magnesium. Increases in boron and sodium may be caused by coolant leakage into the lubricating oil, and increases in silicon may indicate poor filtration of intake air.

Results from short endurance tests of sunflower and cottonseed oils have not indicated abnormal wear patterns or rates for any fuels tested (Engler et al., 1982, 1983b). Although there have been differences in trace metal profiles among the various fuels tested, no general trends have been observed. Some differences between plant oils and diesel fuel may have resulted from dilution effects caused by fuel blowby. Figure 5.17 shows iron profiles for the fuels tested. Greater increases in iron for the plant oils may indicate slightly greater wear with those fuels, but they may also have been caused by blowby, since the plant oils had relatively high iron content (1.5 ppm in sunflower oils).

The most alarming change noted was the increase in the total solids content of the lubricating oil during sunflower oil testing (Figure 5.18). Total solids content exceeding 5 percent is considered dangerous because of possible abrasion to engine parts. All sunflower oils tested caused solids to exceed dangerous levels before the 40-hour test ended; however, solids content did not show similar increases for cottonseed oil fuels.

Longer-Term Testing

In addition to good short-term performance, an acceptable alternative fuel must be able to provide sustained engine operation without

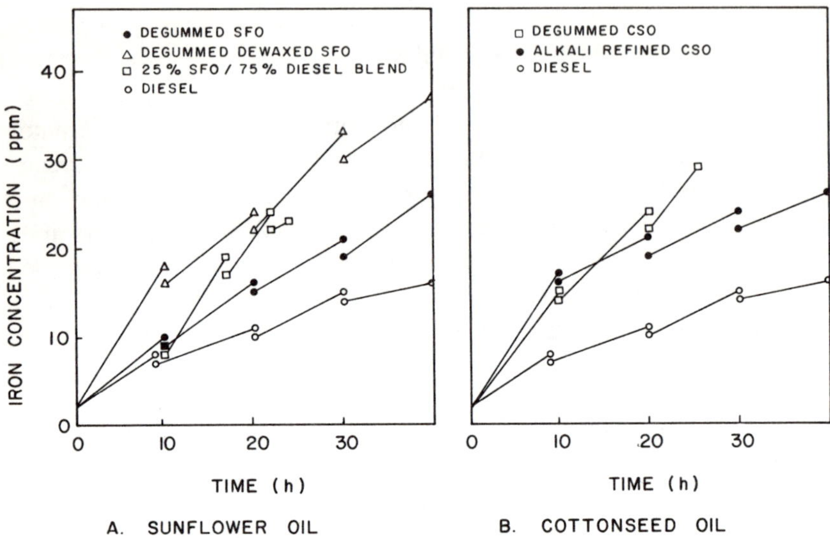

FIG. 5.17. Iron concentrations in lubricating oil samples from short endurance tests using a precombustion-chamber engine.

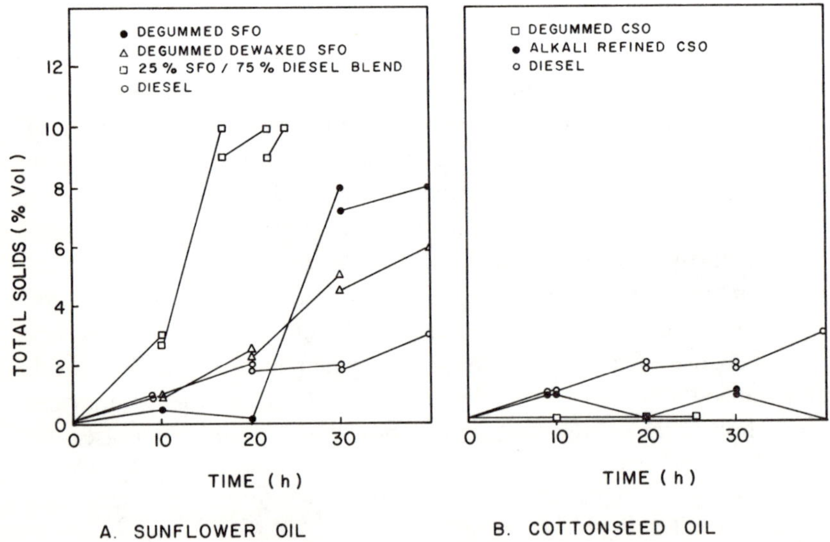

FIG. 5.18. Total solids contents of lubricating oil samples from short endurance tests using a precombustion-chamber engine.

excessive wear or fouling problems. Standardized procedures for fuel screening tests have been developed by the Engine Manufacturers Association (Alternate Fuels Committee, 1982). Reports on the ability of unmodified plant oils to meet these longer-term criteria, although mixed, generally indicate that the use of plant oils—either alone or in blends—leads to increased wear, ring sticking, increased carbon deposits, and more rapid fouling of lubricating oil (Wagner and Peterson, 1982; Walter and Derry, 1982; Ziejewski and Kaufman, 1982). These problems appear to be more severe in direct-injection than in precombustion-chamber engines (Braun and Stephenson, 1982; Bruwer and Hugo, 1981).

Longer-term testing of lower alcohol esters of plant oils indicates that they are good replacements for diesel fuel (Hawkins and Fuls, 1982; Pischinger et al., 1982; Ventura, Noscimento, and Bandel, 1982). The performance of esters is very similar to that of diesel fuel, and there appear to be no deleterious effects on the engine or the lubricating oil. However, injector coking may occur if conversion of plant oil to lower alcohol esters is below 90 percent (Hawkins and Fuls, 1982).

Major Problems

The major problems noted with the use of unmodified plant oils as diesel engine fuels are injector fouling, increased carbon deposits in the combustion chamber, varnish formation on combustion chamber surfaces, ring sticking, cylinder wall scoring, and increased fouling of the lubricating oil. Although these problems may seem diverse, all generally can be traced to the poorer combustion characteristics of plant oil fuels. The properties of plant oils primarily responsible for combustion problems include high viscosity and relatively high contents of unsaturated fatty acids.

Plant oils have viscosities approximately ten times greater than diesel fuel at a typical injector temperature of 40°C (see Table 5.4). These high viscosities alter spray characteristics, causing the formation of larger droplets, which have slower rates of penetration into the combustion chamber and penetrate to a greater distance before they are vaporized (Ryan, Callahan, and Dodge, 1982). Therefore, some plant oil may actually impinge on cylinder walls before it is consumed, resulting in increased blowby into the crankcase or gum formation on the walls. In precombustion-chamber engines, injected fuel is swirled in a vortex of air before entering the combustion chamber. This aids in the atomization of fuel and probably accounts for the better results reported for that type of engine (Braun and Stephenson, 1982; Bruwer and Hugo, 1981).

Plant oil viscosities can be reduced by heating, by blending with diesel fuel or other solvents, by preparing microemulsions, or by interesterification of the oil with lower alcohols. Viscosities of plant oils, ethyl esters, and diesel fuel are shown in Figure 5.19 as functions of temperature. Ryan, Callahan, and Dodge (1982) reported that heating the oil to obtain a viscosity equivalent to that of diesel fuel reduced deposits and other durability problems, but the results obtained by Engler et al. (1983b) showed no reduction in deposits with heated fuels. Viscosity can also be reduced by blending plant oil with diesel fuel, as shown in Figure 5.20. Microemulsions are formed by using an organic surfactant to disperse aqueous ethanol in plant oil; they have viscosities approaching that of diesel fuel. However, microemulsions are expensive, and their effects on durability are unknown at this time (Goering et al., 1982b; Pryde, Schwab, and Freedman, 1983). None of these methods of reducing viscosity affect the chemical properties of plant oils that are also important factors in deposit formation.

Interesterification of plant oils with lower alcohols appears to virtually eliminate combustion problems (Hawkins and Fuls, 1982; Pischinger et al., 1982). The viscosities of methyl and ethyl esters are similar to that of diesel fuel (Table 5.4 and Figure 5.19). In addition, interesterification changes the large branched molecular structure of triglycerides to straight carbon chains of approximately the same length as those found in diesel fuel. Cloud points for methyl and ethyl esters are higher than for diesel fuel, however, and this may cause fuel line plugging and storage problems during cool weather.

The relatively high degree of unsaturation of plant oils is another contributor to combustion problems. Common fatty acid residues found in plant oils are linear chains containing 16 to 18 carbon atoms. These chains may be completely saturated (no double bonds) or contain one or more sites of unsaturation. Typical fatty acid compositions of the major plant oils produced in the U.S. are given in Table 5.1 and over 70 percent of the fatty acids in these oils are unsaturated. Plant oil esters have the same iodine value as the oil from which they were made, since the interesterification reaction does not affect the sites of unsaturation.

Points of unsaturation introduce chemically reactive sites into

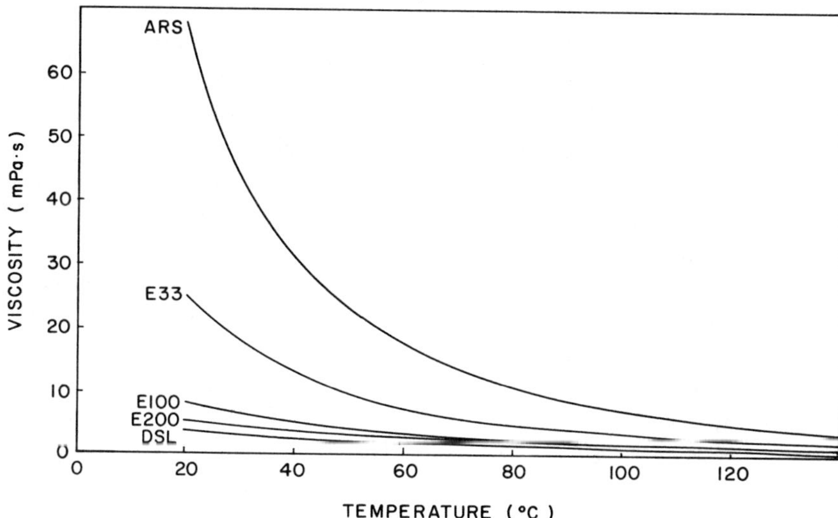

FIG. 5.19. Viscosities of alkali-refined sunflower oil (ARS), sunflower oil ethyl esters (E), and diesel fuel (DSL). Numbers following E denote percentage of stoichiometric amount of ethanol used to prepare ester.

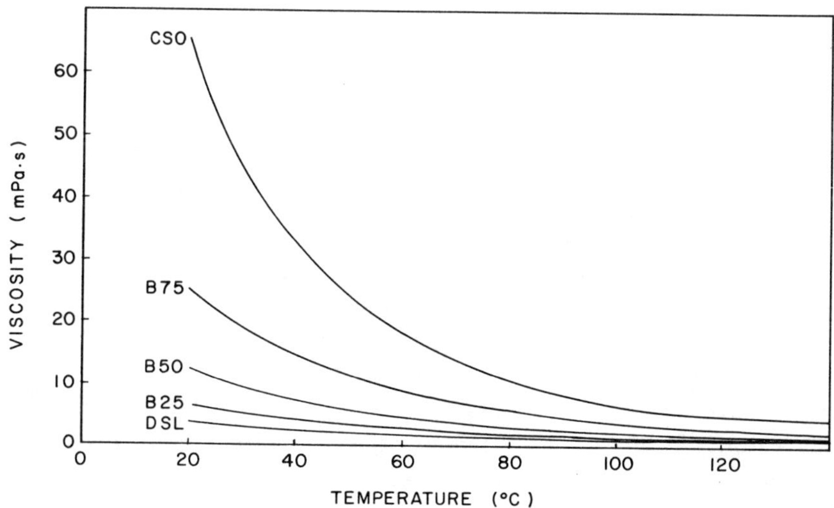

FIG. 5.20. Viscosities of cottonseed oil (CSO) and blends (B) with diesel fuel (DSL). Numbers following B denote percentage by weight of CSO in the blend.

the fatty acid chains, and this reactivity is considerably enhanced in the di- and tri-unsaturated fatty acids. These reactive sites are susceptible to oxidation, which can lead to free radical formation and polymerization reactions with other unsaturated fatty acids. At high temperatures, additional polymerization-causing reactions occur. The mechanisms of both oxidative and thermal polymerization have been reviewed by Formo (1979). Polymerization of triglycerides by either mechanism produces high-molecular-weight, crosslinked gel structures.

In compression ignition engines, the conditions in combustion chambers during injection and combustion of fuel are conducive to both oxidative and thermal polymerization. Large droplets, resulting from poor atomization of plant oils, allow unburned fuel to be subjected to highly reactive conditions for relatively long periods of time. The polymers that form are less likely to be consumed completely by combustion, leading to increased carbon deposits. Contact between unburned plant oils and hot metal surfaces also promotes polymerization and may account for the formation of varnish and deposits that cause ring sticking. The polymerization of unburned fuel (blowby) may be a problem in the crankcase as well. Plant oils are more reactive than diesel fuel and apparently cause more rapid depletion of antioxidants in lubricating oils. Once antioxidants in the crankcase oil have been depleted, polymerization of fuel blowby may cause lubricating oil viscosity to increase and/or the oil filter to become plugged with gums.

Since chemical reactivity increases with unsaturation, oils having high iodine values (>150) would be expected to produce more severe fouling problems. Tests with linseed oil, a highly unsaturated plant oil (IV = 185) used in paints and varnishes because of its ability to form well-developed intertwining polymeric structures, have shown it to cause very rapid injector fouling and loss of power (Quick et al., 1982). Most common plant oils suggested for alternative diesel fuels have iodine values falling in an intermediate range (100–140), and all tend to produce similar fouling problems. Although longer-term engine testing of highly saturated fats such as beef tallow and coconut oil has not been reported, preliminary tests indicate that less fouling occurs with those fuels (Bacon et al., 1981). Unfortunately, highly saturated oils are semisolid at normal ambient temperatures; to be used as fuels, they would require either heated storage facilities, fuel tanks, and fuel lines or dissolution in solvent.

Conversion of linseed oil to methyl esters appears to eliminate fouling problems (Quick, Wilson, and Woodmore 1982). Although the degree of unsaturation remains the same in the methyl esters, the branched structure of triglyceride molecules has been reduced to a smaller straight-chain structure. Oxidative or thermal polymerization may still occur, but it does not proceed past the trimer stage (Formo, 1979) and therefore does not produce three-dimensional gel structures.

Summary

The technical aspects of using plant oils as alternative diesel fuels have been studied extensively during the past few years. Reports generally indicate that thermal efficiencies of straight plant oils range from 90 to 110 percent of that of diesel fuel; power output may decline by up to 5 percent; specific fuel consumption increases by 10 to 15 percent; there are significantly greater amounts of carbon deposits formed on combustion chamber surfaces; and lubricating oil fouls more rapidly. On the other hand, plant oils that have been modified by interesterification with methanol or ethanol appear to be equivalent to diesel fuel in terms of engine performance and fouling problems.

Plant oils contain over 95 percent triglycerides with minor amounts of phosphatides and other components. Non-triglyceride components are removed in various stages of refining. For plant oils to be used directly as alternative fuels, a degumming process to remove phosphatides appears desirable. However, additional refining steps do not appear to improve performance of plant oils as alternative fuels. The fouling problems experienced with straight plant oil fuels appear to be caused by poorer combustion properties and greater reactivity of plant oils toward polymerization. The polymerization of triglycerides causes the formation of a three-dimensional gel structure, which inhibits complete combustion.

Interesterification is a reaction that effectively splits large branched triglyceride molecules into three straight-chain segments. While interesterification does not change reactivity toward polymerization, the resulting polymerization products cannot form a three-dimensional gel network; therefore, combustion can proceed to completion. The conversion of plant oils to methyl or ethyl esters is a relatively simple process, although it does require a high level of quality control for good yields and does not appear suitable for small-scale processing plants.

Degummed or more highly refined plant oils appear to be acceptable fuels for diesel engines on a short-term basis when used with caution: frequent checks of lubricating oil for fouling and periodic inspection of engine cylinders for carbon deposits would be required. Plant oil esters would appear to be even better fuels for temporary use or if economically viable. However, it must be stressed that none of these alternative fuels should be used commercially unless the engine is warranted by the manufacturer for use with that fuel.

References

Adams, C., J. F. Peters, M. C. Rand, B. J. Schroer, and M. C. Ziemke (1983). Investigation of soybean oil as a diesel fuel extender: Endurance tests. *J. Am. Oil Chem. Soc.* 60: 1574–79.

Alternate Fuels Committee (1982). *200-hour screening test for alternate fuels*. Chicago: Engine Manufacturers' Association.

AOCS (1971). *Official and tentative methods*. 3rd ed. Chicago: American Oil Chemists' Society.

Bacon, D. M., F. Brear, I. D. Moncrieff, and K. L. Walker (1981). The use of vegetable oils in straight and modified form as diesel engine fuels. In R. A. Fazzolare and C. B. Smith, eds., *Beyond the energy crisis: Proceedings of the Third International Conference on Energy Use Management*, 3: 1525–33. New York: Pergamon Press.

Braun, D. E., and K. Q. Stephenson (1982). Alternative fuel blends and diesel engine tests. In *Proceedings of the International Conference on Plant and Vegetable Oils as Fuels*, 294–302. St. Joseph, Mich.: American Society of Agricultural Engineers.

Bruwer, J. J., and F. J. C. Hugo (1981). World overview of plant oils for fuel: Status of research and applications technology. In R. A. Fazzolare and C. B. Smith, eds., *Beyond the energy crisis: Proceedings of the Third International Conference on Energy Use Management*, A287–95. New York: Pergamon Press.

Bruwer, J. J., B. V. D. Boshoff, F. J. C. Hugo, J. Fuls, C. Hawkins, and A. N. Vanderwalt (1980). *Sunflower seed oil as an extender for diesel fuel in agricultural tractors*. Silverton, Republic of South Africa: Division of Agricultural Engineering, Department of Agriculture and Fisheries.

Chang, C. C., and S. W. Wan (1947). China's motor fuels from tung oil. *Ind. Eng. Chem.* 39: 1543–48.

Egloff, G., and J. C. Morrell (1932). The cracking of cottonseed oil. *Ind. Eng. Chem.* 24: 1426–27.

Engler, C. R., W. A. LePori, L. A. Johnson, R. C. Griffin, K. C. Diehl,

D. S. Moore, R. D. Lacewell, and C. G. Coble (1982). *Economic and engineering evaluation of plant oils as a diesel fuel*. Final report, Project No. 80-B-4-4a, Texas Energy and Natural Resources Advisory Council, Austin, Tex. College Station, Tex.: Department of Agricultural Engineering, Texas A&M University.

Engler, C. R., L. A. Johnson, W. A. LePori, and C. M. Yarbrough (1983a). Effects of processing and chemical characteristics of plant oils on performance of an indirect-injection diesel engine. *J. Am. Oil Chem. Soc.* 60:1592–96.

——— (1983b). *Engineering evaluation of plant oils as diesel fuel*. Final report, Project No. 80-B-4-4c, Texas Energy and Natural Resources Advisory Council, Austin, Tex. College Station, Tex.: Department of Agricultural Engineering, Texas A&M University.

Formo, M. W. (1979). Paints, varnishes, and related products. In D. Swern, ed., *Bailey's industrial oil and fat products*, 4th ed., 1:687–816. New York: Wiley-Interscience.

Fort, E. F., P. N. Blumberg, H. E. Staph, and J. J. Staudt (1982). Evaluation of cottonseed oils as diesel fuel. Paper (SAE No. 820317) presented at International Congress and Exposition, Detroit, Mich. Sponsored by the Society of Automotive Engineers, Warrendale, Pa.

Freedman, B., and E. H. Pryde (1982). Fatty esters from vegetable oils for use as a diesel fuel. In *Proceedings of the International Conference on Plant and Vegetable Oils as Fuels*, 117–22. St. Joseph, Mich.: American Society of Agricultural Engineers.

Friedlich, J. P., G. R. List, and A. J. Heakin (1982). Petroleum-free extraction of oil from soybeans with supercritical CO_2. *J. Am. Oil Chem. Soc.* 59:288–92.

Goering, C. E., A. W. Schwab, M. J. Daugherty, E. H. Pryde, and A. J. Heakin (1982a). Fuel properties of eleven vegetable oils. *Transactions of the ASAE* 25:1472–77, 1483.

Goering, C. D., R. M. Campion, A. W. Schwab, and E. H. Pryde (1982b). Evaluation of soybean oil–aqueous ethanol microemulsions for diesel engines. In *Proceedings of the International Conference on Plant and Vegetable Oils as Fuels*, 279–86. St. Joseph, Mich.: American Society of Agricultural Engineers.

Graille, J., P. Lozano, P. Geneste, A. Guida, and O. Morin (1981). Production of hydrocarbons by catalytic cracking of palm oil by-products: Development and preliminary tests. *Rev. Franc. Corps. Gras.* 28:421–26.

Grinaker, G. (1981). Oil in them thar fields. *Sunflower* 7 (1): 28–34.

Haag, W. O., P. G. Rudewald, and P. B. Weisz (1980). Catalytic production of aromatics and olefins from plant materials. Paper presented at the American Chemical Society National Meeting, San Francisco, Calif.

Hassett, D. J., and R. A. Hasan (1982). Sunflower oil methyl ester as diesel fuel. In *Proceedings of the International Conference on Plant and Vegetable Oils as Fuels*, 123–26. St. Joseph, Mich.: American Society of Agricultural Engineers.

Hawkins, C. S., and J. Fuls (1982). Comparative combustion studies on various plant oil esters and the long term effects of an ethyl ester on a compression ignition engine. In *Procedings of the International Conference on Plant and Vegetable Oils as Fuels*, 184–97. St. Joseph, Mich.: American Society of Agricultural Engineers.

Hron, R. J., Sr., S. P. Koltun, and A. V. Graci, Jr. (1982). Biorenewable solvents for vegetable oil extraction. *J. Am. Oil Chem. Soc.* 59:674A–84A.

Humke, A. L., and N. J. Barsic (1981). Performance and emissions characteristics of a naturally aspirated diesel engine with vegetable oil fuels (part 2). Paper (SAE No. 810955) presented at International Off-Highway Meeting, Milwaukee, Wis. Sponsored by the Society of Automotive Engineers, Warrendale, Pa.

Johnson, L. A., and E. W. Lusas (1983). Comparison of alternative solvents for oils extraction. *J. Am. Oil Chem. Soc.* 60:181A–94A.

Kusy, P. F. (1982). Transesterification of vegetable oils for fuels. In *Proceedings of the International Conference on Plant and Vegetable Oils as Fuels*, 127–37. St. Joseph, Mich.: American Society of Agricultural Engineers.

Lipinsky, E. S., S. Kresovich, C. K. Wagner, H. R. Applebaum, T. A. McClure, J. L. Otis, and D. A. Trayser (1982). Vegetable oils and animal fats for diesel fuels: A systems study. In *Proceedings of the International Conference on Plant and Vegetable Oils as Fuels*, 1–10. St. Joseph, Mich.: American Society of Agricultural Engineers.

Mangold, H. K. (1983). Liquefied gases and supercritical fluids in oilseed extraction. *J. Am. Oil Chem. Soc.* 60:226–28.

Norris, F. A. (1982a). Extraction of fats and oils. In D. Swern, ed., *Bailey's industrial oil and fat products*, 4th ed., 2:175–251. New York: Wiley-Interscience.

——— 1982b. Refining and bleaching. In D. Swern, ed., *Bailey's industrial oil and fat products*, 4th ed., 2:253–314. New York: Wiley-Interscience.

Peterson, C. L., D. L. Auld, and R. A. Korus (1983). Winter rape oil fuel for diesel engines: Recovery and utilization. *J. Am. Oil Chem. Soc.* 60:1579–87.

Pischinger, G. H., R. W. Siekmann, A. M. Falcon, and F. R. Fernandes (1982). Methylesters of plant oils and diesel fuels, either straight or in blends. In *Proceedings of the International Conference on Plant and Vegetable Oils as Fuels*, 198–208. St. Joseph, Mich.: American Society of Agricultural Engineers.

Pryde, E. H. (1982). Vegetable oil fuel standards. In *Proceedings of the International Conference on Plant and Vegetable Oils as Fuels.* 101–105. St. Joseph, Mich.: American Society of Agricultural Engineers.

Pryde, E. H., A. W. Schwab, and B. Freedman (1983). Oil as fuel. Paper presented at 32nd Oilseed Processing Clinic, New Orleans, La. Sponsored by Southern Regional Research Center, U.S. Department of Agriculture.

Quick, G. R., B. T. Wilson, and P. J. Woodmore (1982). Injector-fouling propensity of certain vegetable oils and derivatives as fuels for diesel engines. In *Proceedings of the International Conference on Plant and Vegetable Oils as Fuels*, 239–46. St. Joseph, Mich.: American Society of Agricultural Engineers.

Rhee, K. C., C. M. Cater, and K. F. Mattil (1972). Simultaneous recovery of protein and oil from raw peanuts in an aqueous system. *J. Food Sci.* 37:90–93.

Romano, S. (1982). Vegetable oils: A new alternative. In *Proceedings of the International Conference on Plant and Vegetable Oils as Fuels*, 106–16. St. Joseph, Mich.: American Society of Agricultural Engineers.

Ryan, T. W. III, T. J. Callahan, and L. G. Dodge (1982). Characterization of vegetable oils for use as fuels in diesel engines. In *Proceedings of the International Conference on Plant and Vegetable Oils as Fuels*, 70–81. St. Joseph, Mich.: American Society of Agricultural Engineers.

Sonntag, N. O. V. (1979). Composition and characteristics of individual fats and oils. In D. Swern, ed., Bailey's industrial oil and fat products, 4th ed., 1:289–477. New York: Wiley-Interscience.

Sonntag, N. O. V. (1982). Fat splitting, esterification and interesterification. In D. Swern, ed., Bailey's industrial oil and fat products, 4th ed., 2:97–173. New York: Wiley-Interscience.

Strayer, R. C., J. A. Blake, and W. K. Craig (1983). Canola and high erucic rapeseed oil as substitutes for diesel fuel: Preliminary tests. *J. Am. Oil Chem. Soc.* 60:1587–92.

Ventura, L. M., A. C. Noscimento, and W. Bandel (1982). First results with Mercedes-Benz DI diesel engines running on monoesters of vegetable oils. In *Proceedings of the International Conference on Plant and Vegetable Oils as Fuels*, 394–400. St. Joseph, Mich.: American Society of Agricultural Engineers.

Vinyard, S., E. S. Renoll, J. S. Goodling, L. Hawkins, and R. C. Bunt (1982). Properties and performance testing with blends of biomass alcohols, vegetable oils and diesel fuel. In *Proceedings of the International Conference on Plant and Vegetable Oils as Fuels*, 287–93. St. Joseph, Mich.: American Society of Agricultural Engineers.

Wagner, G. L., and C. L. Peterson (1982). Performance of winter rape (*Brassica napus*) based fuel mixtures in diesel engines. In *Proceedings of the*

International Conference on Plant and Vegetable Oils as Fuels, 329–36. St. Joseph, Mich.: American Society of Agricultural Engineers.

Walter, J., P. Aakre, and J. Derry (1982). The 1981 "flower power" field testing program. In *Proceedings of the International Conference on Plant and Vegetable Oils as Fuels*, 384–93. St. Joseph, Mich.: American Society of Agricultural Engineers.

Weisz, P. B., W. O. Haag, and P. G. Rudewald (1979). Catalytic production of high-grade fuel (gasoline) from biomass compounds by shape-selective catalysts. *Science* 206:57–58.

Wiebe, R., and J. Nowakowska (1949). The technical literature of agricultural motor fuels. *USDA Bibliographical Bulletin* 10:52–105, 183–95.

Yarbrough, C. M., W. A. LePori, and C. L. Darcey (1982). Cottonseed oil as a diesel fuel substitute. Paper (No. 82-3613) presented at American Society of Agricultural Engineers winter meeting, Chicago, Ill.

Ziejewski, M., and K. R. Kaufman (1982). Laboratory endurance test of a sunflower oil blend in a diesel engine. In *Proceedings of the International Conference on Plant and Vegetable Oils as Fuels*, 354–63. St. Joseph, Mich.: American Society of Agricultural Engineers.

CHAPTER 6

Systems Engineering of Biomass-Fueled Energy Alternatives

CALVIN B. PARNELL, JR.

The goal of the ongoing research in developing technology to transform biomass into usable energy is the implementation of this research. Applications of biomass conversion technology are attractive because such fuel is a renewable natural resource and not subject to international politics. However, the conversion of biomass into usable energies is plagued with such uncertainties as economic feasibility, environmental regulations, fuel supply, and technical labor requirements. The use of systems engineering in the analysis and design of a biomass-fueled energy alternative can reduce uncertainties and increase the probability of success.

The use of agricultural waste as a fuel for electric power generation or cogeneration has been the subject of the simulation studies that are discussed in this chapter. Can small power plants be built at cotton gins, cottonseed oil mills, or textile mills to generate electricity and offset all or a portion of the electric power purchased? How long should the power plant operate to be economical? How much fuel would be available? What size power plant would be optimal? What would be the economic benefit of recovering waste heat or steam? Would it be simpler or more economical if agricultural waste were to be marketed to a coal-fired power plant as a fuel extender?

If biomass is to have an impact on our future agricultural energy requirements, new concepts will have to be considered. Electric power is currently purchased from large, centralized generation systems. Suppose that in the future many small decentralized power plants (<10 MW) fueled with agricultural waste were utilized. Recent federal regulations (Federal Energy Regulatory Commission, 1978, 1980) have provided the incentive for development of small-power generation. According to these regulations, a small plant could furnish electrical power to an on-site processing plant or large farm and sell its excess electricity back to the power company at the company's "avoided cost."

It would be desirable to have biomass conversion systems operating today that would significantly reduce our dependence upon fossil fuels, especially foreign oil. How much energy can we expect to derive from biomass? Good (1981) estimated that the upper limit of annual photosynthetic conservation of solar energy in the United States was 106 exajoules, but that practical limitations of accessibility, as well as high transportation and conversion costs, would limit the net energy gain to 4 exajoules. The U.S. consumed 80.2 exajoules in 1980, and Gustaferro (1981) predicted that it will consume 102 exajoules in the year 2000. A net energy gain from photosynthesis of 4 exajoules is 5 percent of the energy consumed in 1980. To illustrate the magnitude of U.S. dependence on fossil fuels: in 1978 we produced a total of 610 million tonnes of agricultural products and consumed 1,960 million tonnes of fossil fuels (Gustaferro, 1981). Hence, it is not likely that biomass-derived usable energy will displace large volumes of fossil fuels, barring major breakthroughs in research.

How large should a biomass conversion system be? Scale is a central factor in design. As mentioned in Chapter 3, small-scale alcohol plants (less than two million gallons per year) are not likely to be economical. A 20-MW power plant is small in comparison to a 500-MW utility; however, this same 20-MW power plant would seem very large to a cotton ginner or an oil mill operator. For the purpose of discussion, biomass conversion system sizes will be defined as follows:

1. A large conversion system is any system larger than 80 MW that produces more than 290 gigajoules per hour of steam or gas energy and exceeds 3,500 gallons of alcohol per hour at 80 megajoules per gallon.

2. A medium conversion system will produce less usable energy than a large system but more than a small one.

3. A small conversion system is any system smaller than 10 MW that produces less than 36 gigajoules per hour of steam or gas energy and less than 450 gallons per hour of alcohol at 80 megajoules per gallon.

In developed nations, public expectation of shortages and high prices has stimulated the development of new technology in the hope that economically competitive alternatives will be immediately available when needed. One of the most critical needs will be the energy required by agriculture for the production of food and fiber. It would seem logical that one of our immediate goals should be to strive toward use of agricultural wastes in biomass conversion systems so that agricultural production and processing can be less dependent upon fossil fuels. The National Academy of Sciences (1979) estimated that we could displace 0.53 exajoules of fossil fuel energy in 1985 by using agricultural wastes, and 3.7 exajoules in the year 2000. Gavett (1980) predicted that we would require 2.33 exajoules for agricultural production and processing in 1990. This estimate included the production of 38 gigaliters of ethanol.

The concept of reducing agriculture's dependence on energy derived from fossil fuels would seem to have merit. With a potential of 3.7 to 4 exajoules of available energy from agricultural residues and a projected need for 2.3 exajoules in the year 1990, one would think that the prospects for this goal would be bright. However, agriculture, unlike much of our energy industry, can be characterized as largely decentralized. The implementation of any biomass conversion system that does not have a high probability of economic success is not likely. Hence, if biomass conversion systems are to be adapted to meet agriculture's energy needs, they will likely be small in scale to reduce the initial investment, and decentralized so that they can be located at or near the source of fuel (agricultural waste). In addition, the design of each system will be site-specific, requiring considerable engineering. The engineering associated with these conversion systems will have to cope with relatively poor thermal conversion efficiencies, waste disposal, and an industry that is not accustomed to hiring technical labor.

How should small-scale, decentralized, biomass conversion systems—fueled with agricultural wastes—be designed? The problem is not simple. The new system must integrate into the existing energy supply systems and displace the energy derived from fossil fuels in a stepwise manner. Difficulties such as the changing costs of biomass fuels, the reliability (downtime) of conversion systems, technical labor requirements, compliance with environmental regulations, waste disposal, and safety must be anticipated. Above all, every effort must be made to insure that a high probability of success exists prior to the purchase of hardware. The objective of a biomass-fueled energy alternative should be a system that significantly reduces dependence on energy derived from fossil fuels and is at the same time technically and economically feasible.

Systems Engineering and Simulation

The design and implementation of biomass conversion technology is an ideal application for systems engineering. Systems engineering can be defined as a combination of engineering, engineering design, and operations research. A systems engineering approach to designing a biomass conversion facility would comprise the following features:

— a "systems" point of view;
— matching of inputs and outputs of various components;
— consideration of fuel supply and energy needs in sizing and locating plant;
— site-specific feasibility studies;
— some degree of optimization;
— simulation and model testing to include a macroeconomic analysis.

The use of agricultural wastes to displace fossil fuel energy offers significant potential for agricultural processing plants in the immediate future. Agricultural processing plants are usually located in areas where access to large volumes of inexpensive biomass is not a problem. Cotton gins, oil mills, and textile mills have experienced dramatic increases in expenditures for steam, heat, and electricity. Electricity is the most expensive energy derived from fossil fuels, but the price of natural gas, which is used to produce heat and steam, has also increased greatly. One of the most promising alternatives is a small power cogeneration system fueled with agricultural biomass waste to provide (1) electricity for cotton ginning with waste heat recovery for seed cotton drying, (2) electricity for cottonseed oil mill processing with waste steam recovery for process requirements, or (3) electricity for textile mill processing with waste steam recovery for process requirements.

The major variables that will affect the success of such applications are (1) power plant size, (2) sophistication of heat and steam recovery design, (3) fuel costs, (4) fuel supply, (5) the technology used (gasification or combustion), (6) system reliability, (7) labor requirements, (8) environmental and safety constraints, and (9) the cost/benefit of usable energy derived from biomass conversion as compared to conventional energy sources.

Figure 6.1 is a typical systems engineering approach to the design of a small power cogeneration system for one of the applications de-

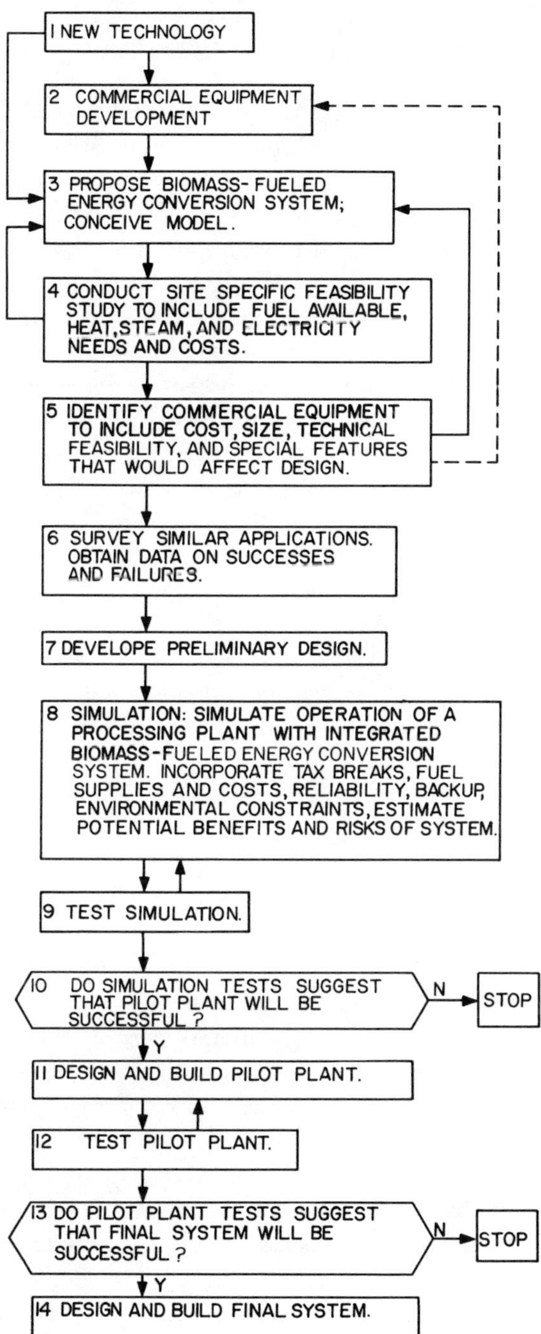

FIG. 6.1. Steps of a systems engineering approach to the design and successful implementation of a biomass-fueled energy alternative.

scribed above. The first seven steps would normally be incorporated in any engineering approach to a preliminary design. Steps 8 and 9 (simulation and testing) are unique to a systems engineering approach. A good simulation model can be used to detect bottlenecks, design flaws, and critical variables. But most important, simulation and testing can be used to estimate the resulting cost/benefit of the biomass conversion system. If results of the simulation suggest that the system will not be successful, it should be stopped or the system should be redesigned. Simulation testing will demonstrate the effects of critical variables on the potential success of the system's design. If sufficient reliable data are not available for these critical variables, steps 11, 12, and 13 should be considered prior to building the final system.

Design of biomass energy conversion systems is currently in the forefront of technology. It should be recognized that problems will occur during the simulation and pilot-plant phases. The systems engineering approach is dynamic in that successive steps will feed back information to the preceding steps, thus continually updating the design specifications for the implementation of this technology.

The benefits of formulating and testing a computer simulation model are as follows:

1. The process of formulating the simulation model can lead to a better understanding of the proposed system design, which can lead to design improvements.

2. Many of the critical design decisions can be tested, and the results can assist in improving the systems design.

3. Simulation allows for evaluation of the effects of complex component interactions upon a system's performance.

4. Conducting a series of experiments permits the identification of critical design variables and potential bottlenecks.

5. Simulation testing enables the study to proceed in simulated real time.

The two examples included in this chapter demonstrate the kind of information that can be provided by simulation and testing. The first is a small power plant at a cottonseed oil mill. The electric power generated could be used by the plant to reduce or replace electric power purchased from the utility company. The second would seem to be attractive for large-scale agricultural biomass energy conversion: using cotton gin trash as a fossil fuel extender in a coal-fired power plant. In this case the consumer of the biomass would be the electric power utility, and large volumes of biomass would be needed.

Biomass Fuels for Small Power Plants or Cogeneration

As a consequence of recent federal legislation (FERC, 1978, 1980), incentives exist for private industry to seriously consider cogeneration or small-scale electrical power production derived from renewable natural resources. Allison (1981) listed these incentives as follows: (1) a total of 20 percent investment and energy tax credits, (2) an assured market for independently generated power at a just and reasonable rate, (3) exemption from federal and state public utility regulations, (4) exemption from incremental gas pricing, and (5) accelerated depreciation schedules.

To be considered a small power producer, an agricultural processing facility (1) must produce no more than 80 megawatts of power, (2) must derive no more than 25 percent of the total energy input from oil or natural gas, (3) must generate 75 percent of the electrical power from renewable fuel resources such as biomass waste, and (4) must not be owned by a person primarily engaged in generation or sale of electric power (Public Utility Commission of Texas, 1983a).

The passage of the Public Utility Regulatory Policies Act (PURPA) in 1978 and the Section 210 rules in 1980 by the Federal Energy Regulatory Commission (1978, 1980) was intended to encourage development of cogeneration or small power production. Cogeneration is viewed as the sequential production of electricity and thermal energy (heat or steam) from a single heat source, and small power production is the production of electricity from renewable resources. The relative success of the act in Texas is illustrated by the construction of 68 new small power plants (Table 6.1) following the enactment of these new federal regulations (PUC, 1983b). However, 67 of these new plants were wind generators; the uncertainty associated with the engineering design and successful operation of small power plants fueled with agricultural wastes has had an impact on the acceptance of this potential source of usable energy. In addition, questions associated with cost of backup power, demand charges, "avoided cost" rates, and reliability must be addressed by the state's public utility commission.

Cotton gins and cottonseed oil mills are paying high prices for energy. The cost of electrical energy has increased from more than $0.02 per kWh in 1972 to more than $0.10 per kWh at some locations in 1983, with an associated tripling and quadrupling of natural gas and propane prices. Texas produces and processes an average of four million bales of cotton each year. Approximately 90 percent of this is

TABLE 6.1 Small Power Producers and Cogenerators in Texas (1982)

	Existing Facilities[(a)]	New Facilities[(b)]	Total Facilities	Existing Capacity (kW)[(a)]	New Capacity (kW)[(b)]	Total Capacity (kW)
Small Power Producers	1	68	69	5,000	1,113	6,113
Cogenerators	20	6	26	1,303,500	286,000	1,589,500

Source: Public Utility Commission of Texas (1983b).
(a) Construction commenced before 11/9/78 (before PURPA).
(b) Construction commenced on or after 11/9/78 (after PURPA).

stripper-harvested cotton with each bale containing approximately 225 kg of lint fiber, 410 kg of seed, and 450 kg of sticks, burs, motes, and fine trash, i.e., gin trash (Baker, 1977). The thermal energy content of this resource is 14 to 16 MJ/kg with an ash content of 15 to 20 percent. More than 1.6 megatons of gin trash is available to serve as fuel for small power plants or cogeneration. Gin trash has an added advantage over corn stover, wheat straw, and sorghum stubble in that it has already been harvested and delivered to a central location (the gin). Conventional disposal of this waste product consists of land application by means of spreader trucks, resulting in a cost to the ginner of $6.50 to $10 per tonne (Parnell, 1983).

Cottonseed oil mills in Texas process in excess of 1.2 megatons cottonseed each year containing 240 kilotons of hulls. Approximately 10 percent of the hulls are used by the mill process to reduce the protein content of cottonseed meal to 41 percent for marketing purposes. The remainder of the hulls are sold as cattle feed with a highly variable market value. At times, the net return for hulls is less than the cost of transportation and storage. Hulls are an attractive biomass fuel, having a relatively high thermal energy content of 16 to 18 MJ/kg and a low ash content (3–4 percent).

The oil mill process is energy intensive, requiring 130 to 200 kWh of electrical energy and 0.5 to 1 GJ of thermal energy for steam generation per tonne of cotton seed processed. Oil mills are located in cotton-producing areas near cotton gins. Some oil mills have "line" gins: gins owned and managed by the oil mill organization. Hence, the use of gin trash as a fuel for a small power plant or cogeneration system at an oil mill would be possible, and may be preferable in years when the market value of hulls is favorable. Like cotton gin trash, hulls accumulate and are stored on the site of the potential biomass-fueled energy conversion system.

If cotton gin trash were to be used at a cotton gin, oil mill, or centralized power plant as a biomass fuel, a new materials-handling system would be required to condense, store, and retrieve this resource. Parnell (1983) conceived of using the cotton module builder and transporter for this purpose. The gin trash moduling process would be conducted simultaneously with ginning, and the resulting nine-tonne modules removed to a storage location by means of a module transporter. The advantage of this process is that the technology is already being used to handle seed cotton. Strother (1982) used this concept successfully in transporting gin trash to cattle feedlots.

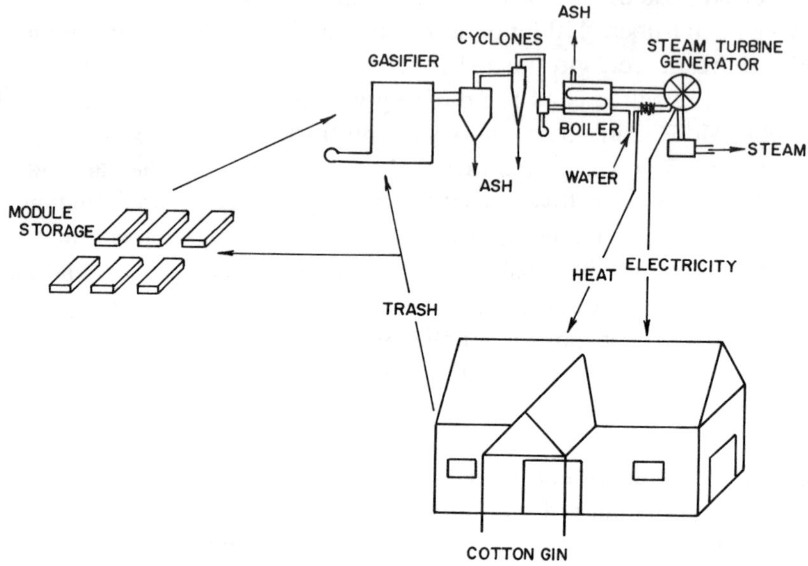

FIG. 6.2. Systems concept of a small power plant, fueled with agricultural wastes (gin trash), generating electricity and heat for a cotton gin.

The systems concepts shown in Figures 6.2 and 6.3 are conceived with a gasification thermal conversion system as the critical component. As described in Chapter 2, gasification is the most promising thermal conversion technology for cotton gin trash. Fluidized-bed gasification provides for thermal reactions at controlled temperatures (<800°C) to prevent slagging and fouling. Reduced gas volume associated with gasification simplifies and significantly reduces the cost of ash removal, and reduces or eliminates the metal corrosion that occurs when gin trash is burned in an excess-oxygen environment.

Figure 6.4 depicts the most promising biomass thermal conversion system; it is scheduled for testing in 1984 (LePori et al., 1983; Parnell et al., 1984). It consists of a feeder, fluidized-bed gasifier, ash removal system, and boiler. Fuel is fed to a fluidized-bed gasifier, resulting in a low-energy gas (4.0–6.0 MJ/m^3 [std]). The ash is carried with the gas to an ash removal system consisting of cyclones in series. The resulting clean gas is conveyed to a boiler to produce steam. Fabric filtration is shown collecting the particulate emissions of the spent fuel to meet community air pollution regulations.

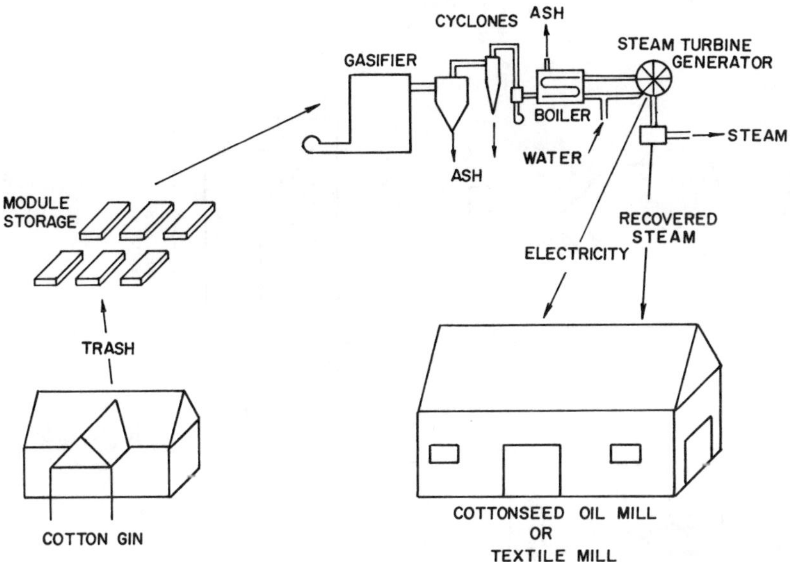

FIG. 6.3. Systems concept of a small power plant fueled with agricultural wastes (gin trash or cottonseed hulls) generating electricity and steam.

Datin (1983) found that 92 to 96 percent of the ash can be removed by a cyclone-series system. The concentration of particulate remaining in the cleaned gas was 2 to 4 g/m^3 (std). However, more than 50 percent of the particulate remaining in the gas stream leaving the ash removal system was carbon, adding to the gas energy content. The effectiveness of the ash removal system is critical to the success of the biomass conversion system illustrated in Figure 6.4. If insufficient ash is removed, slagging, fouling, and metal corrosion are likely to occur in the high-temperature zones of the combustion chamber and boiler. The benefits of fluidized-bed gasification are dependent upon effective ash removal.

Figure 6.5 is a schematic of typical energy consumption and losses that are likely to occur in a small power plant fueled with cotton gin waste. The thermal conversion efficiencies correspond to a design that includes a fluidized-bed gasifier, ash removal system, fire-tube boiler, and steam turbine generator. Fire-tube boilers are not normally used to produce steam at pressures of more than 1 kPa. The thermal conversion efficiency of a steam turbine generator is highly dependent

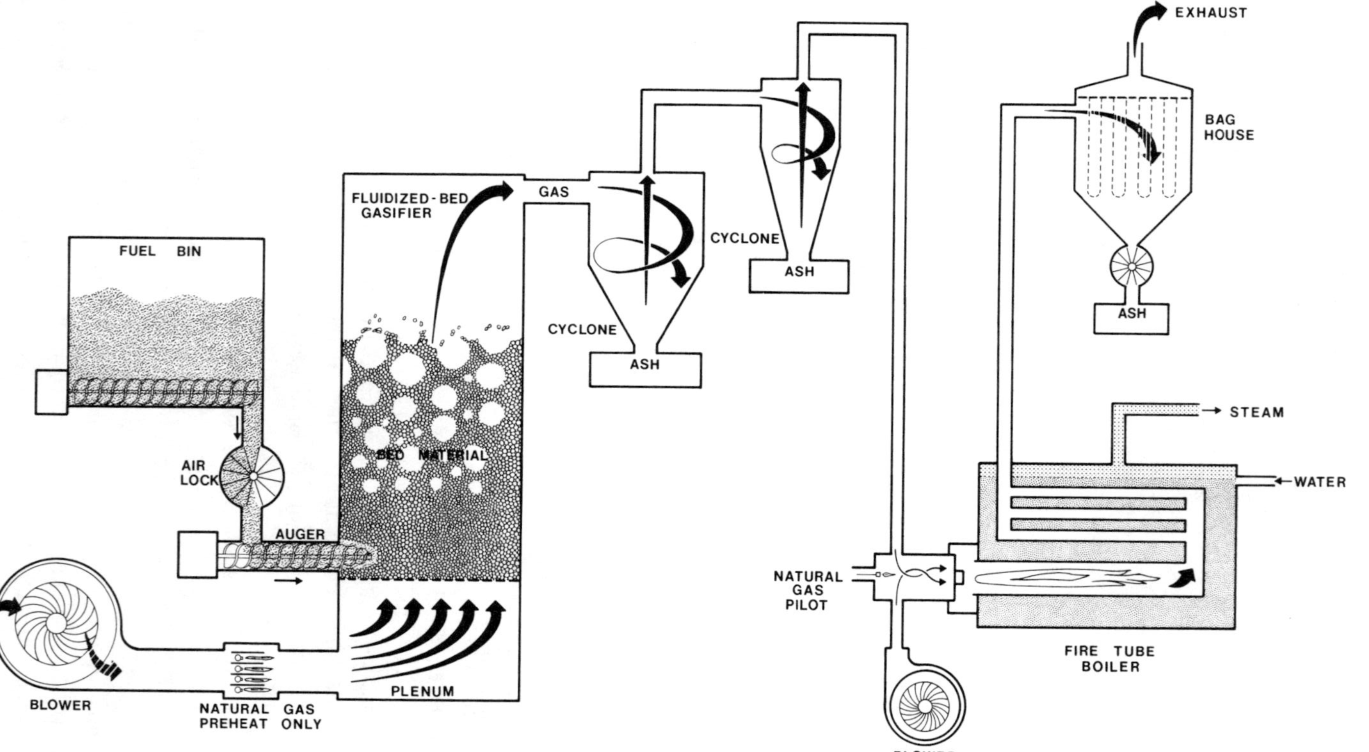

FIG. 6.4. Texas A&M Research fluidized-bed gasifier boiler II.

on the steam pressure and turbine design. At 1 kPa steam pressure, turbine generators can achieve thermal conversion efficiencies of 12 to 16 percent. The 12 percent thermal conversion efficiency can be obtained with a single-stage turbine at a cost of $120,000 (1983 dollars); the 16 percent conversion efficiency would require a multistage turbine at a cost of $193,000 (Worthington Division, McGraw Edison Company, Houston, Texas).

The design illustrated in Figure 6.5 includes a steam turbine with a 15 percent thermal conversion efficiency. The overall thermal conversion efficiency of the design is 8.1 percent. For this system, 2,700 kg/h of fuel would be required to produce 1 MW of electrical power. If the design were to be changed to include a water-tube boiler capable of producing high-pressure steam, a steam turbine generator could be obtained with a thermal conversion efficiency of 25 percent. This change in engineering design would reduce fuel requirements to 1,700 kg/h with a resulting net reduction of 1 tonne/h of fuel. The initial cost of this design, however, would be much greater.

If fuel supplies are inexpensive and plentiful, it may be possible to build and operate a less costly, even though less efficient, power plant. Fuel supplies and fuel costs are factors in the economical production of power that must be considered in the formulation of a design. If the cost of building the most efficient small power plant is prohibitive, then a design that incorporates equipment with intermediate efficiencies may be optimal.

Simulation of Biomass-Fueled Electric Power Generation at a Cottonseed Oil Mill

Designing a small power plant for a cottonseed oil mill involves a number of questions, many of which cannot be answered accurately until the power plant has been constructed and is on line. However, considerable insight can be gained by an engineer with a well-conceived simulation model.

What would be the optimal size? A simple answer would be to size the power plant to supply all the electric power required by the oil mill. An alternative would be to size the power plant on the basis of the reliable residue fuel available. For a cottonseed oil mill, a logical choice of fuel would be cottonseed hulls. Approximately 200 kg of hulls are produced from every tonne of seed processed. Of this total, 10 percent is utilized to reduce the meal protein to 41 percent for mar-

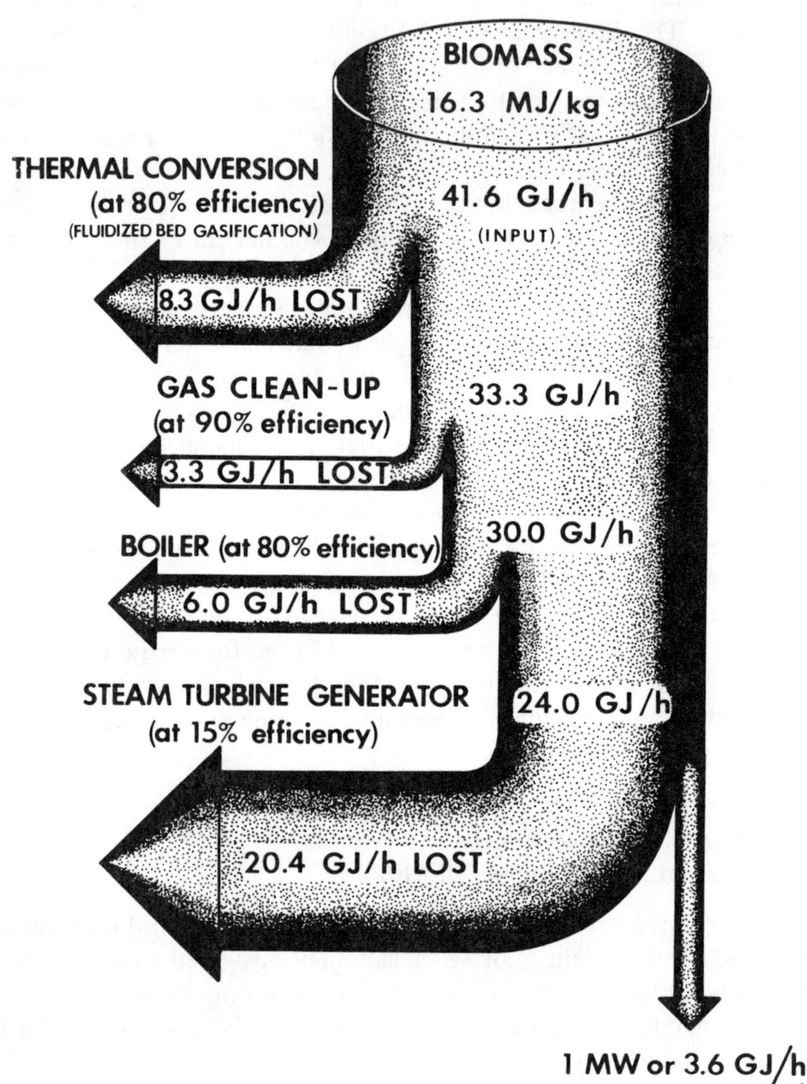

FIG. 6.5. Energy flow in proposed small power plant.

keting purposes. Conceivably, a power plant could be sized to operate on a fuel supply consisting of 80 percent of hull production and never be without fuel while the oil mill operates.

A simulation model was formulated to depict the operation of a cottonseed oil mill and small power plant. The model was used to test the hypothesis that there would be sufficient electrical energy from a power plant sized on 80 percent of hull production to power the delintering room, which is one of the most energy-intensive areas in an oil mill. The model was coded in the SLAM (Simulation Language for Alternative Modeling) language (Pritsker and Pegden, 1979). Four sizes of oil mills (200, 300, 400, and 500 tonnes/day) were studied. Figure 6.6 is a flow chart depicting the simulated system. The assumptions made for simulation testing were as follows:

(a) that the net cost of cottonseed hulls was \$10/tonne, \$15/tonne, \$20/tonne, and \$25/tonne;
(b) that the gasifier was designed on the basis of 15 GJ/h per square meter of bed area;
(c) that the power plant would be down 5 to 15 percent of the operating time, requiring the purchase of electricity at \$0.08/kWh;
(d) that ash disposal costs were \$6.50/tonne;
(e) that the ash content of the fuel was 15 percent;
(f) that fuel energy content was 16 MJ/kg.

Typical simulation input data are shown in Table 6.2. The three oil mills processed 300, 400, and 500 tonnes/day and were operated for 330 days/year. The power plants, sized on the basis of a fuel rate of 80 percent of hull production, were of 1.6, 2.1, and 2.6 MW capacities. Fuel cost was \$10 per tonne. The gasifier design was based on 15 GJ/h per square meter of bed area per hour. Downtime was estimated to be 5 percent of the operating time. Table 6.3 illustrates typical results produced by this simulation. The resulting cost of electrical power was \$0.04 to \$0.053 per kWh.

Mill utilization is defined as the average annual processing rate divided by the theoretical maximum. For example, a 500-tonne/d oil mill will average 400 tonne/d at 80 percent utilization. Table 6.4 summarizes the total hull electricity, and ash production per year for the four oil mill sizes operating at 100, 80, 60, and 40 percent capacity. Table 6.5 lists the size of power plant, critical cost, and yearly excess electricity produced over that needed for the linter room.

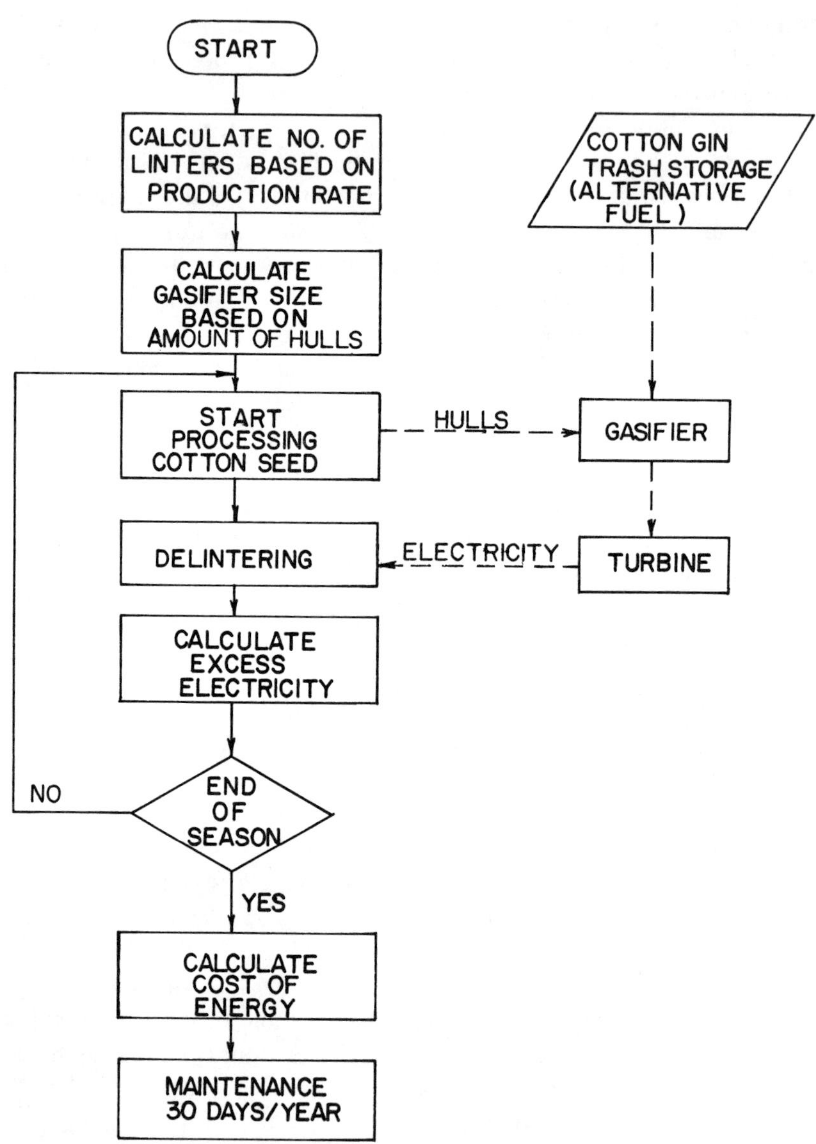

FIG. 6.6. Schematic of SLAM simulation program for a cottonseed oil mill with a small electric power plant.

TABLE 6.2 Typical Inputs for Computer Simulations of Three Operations

	No. 1	No. 2	No. 3
Mill characteristics			
Mill size (tonnes/day)	300	400	500
Mill season (days)	330	330	330
Number of first-cut linters	8	11	13
Number of second-cut linters	24	33	39
Cost of hulls ($/tonne)	10	10	10
Gasifier characteristics			
Input consumption of gasifier ($GJ/m^3/h$)	15	15	15
Diameter of gasifier (meters)	2	2	2
Input rate of solid fuel (kg/h)	2,000	2,670	3,330
Input rate of air (liters/second)	745	993	1,240
Fuel to air ratio	.61	.61	.61
Conversion efficiency of gasifier (%)	70	70	70
Conversion efficiency of boiler (%)	80	80	80
Conversion efficiency of turbine (%)	30	30	30
Expected amount of ash (kg/h)	300	400	500
Size of air blower motor (kW)	13.4	17.9	22.4
Estimated size of turbine (MW)	1.6	2.1	2.6
Expected total kW in lintering room	845	1160	1370
Percent utilization of hulls	80	80	80
Expected life of equipment (yr.)	5	5	5
Current electricity rate ($/kWh)	0.08	0.08	0.08

The use of simulation models provides an opportunity to conduct sensitivity analyses of specific design variables. For example, biomass fuel costs can vary greatly. Figure 6.7 demonstrates how the cost of electricity will be affected by fuel costs ranging between $10 and $25 per tonne for 100, 80, 60, and 40 percent utilization rates. For example, at a processing rate of 400 tonnes per day (80 percent) for a 500-tonne-per-day oil mill, the cost of electricity will range between $0.047 and $0.067 per kWh.

One of the most difficult variables to quantify in system designs for new technology application is the downtime. Figure 6.8 provides cost-of-electricity results for a 200-tonne-per-day oil mill operating at 100, 80, 60, and 40 percent utilization rates as a function of percent of downtime. At a processing rate of 160 tonnes per day (80 percent), the

TABLE 6.3 Typical End-of-Season Results from Computer Simulations

	No. 1	No. 2	No. 3
Capital costs			
Estimated cost of gasifier ($1,000)	510	550	550
Estimated cost of burner ($1,000)	100	100	120
Estimated cost of boiler ($1,000)	100	100	100
Estimated cost of turbine ($1,000)	315	420	525
Total ($1,000)	1,025	1,170	1,295
Fixed costs			
Yearly cost of gasifier ($1,000)	105	113	113
Yearly cost of low-Btu burner ($1,000)	21	21	25
Yearly cost of boiler ($1,000)	21	21	21
Yearly cost of turbine ($1,000)	65	86	108
Total ($1,000)	212	241	267
Variable costs			
Cost of electricity ($1,000)	.18	.24	.29
Cost of hulls ($1,000)	158	211	263
Fuel ash transportation ($1,000)	15	20	26
Maintenance cost ($1,000)	102	117	129
Salary of supervisors ($1,000)	149	149	149
Salary of extra labor ($1,000)	74	74	74
Total ($1,000)	498	571	641
Benefits			
Income from excess hulls ($1,000)	40	53	66
Expected tax credit for equipment ($1,000)	22	26	28
Power plant report			
Total electricity produced (MWh)	12,200	16,200	20,300
Excess electricity produced (MWh)	5,390	6,910	9,260
Cost ($/kWh)	.053	.045	.040

production costs increase from $0.069 to $0.093 per kWh. Increasing the power system's expected life from five to ten years, for the same 200-tonne-per-day oil mill operating at 80 percent capacity, reduces the cost of electricity production from $0.084 to $0.074 per kWh (Figure 6.9).

A key variable used in sizing the fluidized-bed gasifier is the biomass fuel rate per unit of bed area, expressed in units of gigajoules per square meter per hour. Some manufacturers have indicated that they can design a gasifier based on 23 $GJ/m^2/h$. LePori (1982) has found

TABLE 6.4 Hulls, Electricity, and Ash from Power Plant Fueled by Cottonseed Hulls

Rated Capacity (tonnes/day)	Mill Utilization 100%	80%	60%	40%
	Total Hulls Used per Year (tonnes)			
500	32,970	26,350	19,770	13,200
400	26,370	21,090	15,820	10,560
300	19,790	15,830	11,870	7,915
200	13,204	10,550	7,912	5,275
	Total Electricity Generated per Year (MW·h)			
500	20,300	16,290	12,190	8,100
400	16,240	13,030	9,753	6,482
300	12,180	9,774	7,316	4,860
200	8,126	6,517	4,877	3,239
	Total Ash per Year[(a)] (tonnes)			
500	3,940	3,150	2,374	1,584
400	3,153	2,522	1,899	1,268
300	2,364	1,892	1,425	951
200	1,577	1,262	951	634

(a) Ash content = 15%.

that 11.5 GJ/m^2/h as the design criterion will yield a workable gasifier. Sensitivity analysis can evaluate the impact of the engineering design decision on the final cost of electricity production for an oil mill and power plant of any size.

If the assumptions associated with these simulations are correct, a 500-tonne-per-day oil mill operating at 100 percent capacity could produce electricity at a cost of less than $0.05/kWh. The size of the power plant would be less than 3 MW. The resulting cost suggests that this application of biomass conversion technology has promise.

Simulation of Supplemental Biomass Fuel for a Conventional Power Plant

Cotton gin trash has an energy content of approximately 16 MJ per kilogram, which is similar to the energy content of low-grade coal.

TABLE 6.5 Size and Cost of Power Generating System, and Excess Electricity Produced

Rated Capacity (tonnes/day)	Generating Size[a] (kW)	System Cost[b] ($1,000)	Excess Electricity Produced per Year[c]			
			100% (MWh)	80% (MWh)	60% (MWh)	40% (MWh)
500	2,624	1,300	9,264	5,250	1,270	0.3[d]
400	2,100	1,170	6,908	3,698	655	0.0
300	1,574	1,030	5,391	2,984	636	0.2
200	1,050	900	3,876	2,270	653	0.4

(a) Based on 80% of anticipated hull production at 100% of mill utilization.
(b) Power plant costs based on sizing gasifier for 15.0 GJ/m^2h.
(c) Beyond what is needed for linter room. Includes 5% downtime for power generation system.
(d) Negative numbers indicate net deficit for linter room.

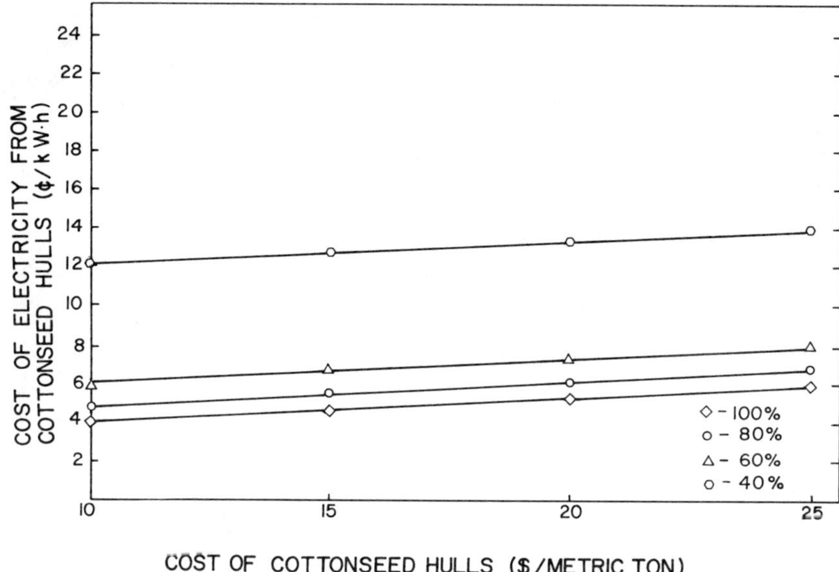

FIG. 6.7. Cost of electricity generation for a 500 tonne-per-day oil mill operating 330 days per year, 7 days per week, 24 hours per day fueled by cottonseed hulls at \$10, \$15, \$20, and \$25 per tonne for 100, 80, 60, and 40% utilization rate.

The concept of using gin waste in a coal-fired power plant is an example of large-scale utilization and would require that large volumes of this biomass be available in the vicinity of the plant. If only 7 percent of the thermal input of a 500-megawatt coal-fueled power plant were supplied with cotton gin waste, the utilization rate would be 27.3 tonnes (three modules) per hour, for 24 hours per day, seven-days per week. Hence, in 100 days, 65.3×10^3 tonnes of gin waste would be required. A SLAM simulation model was written to study the feasibility of using cotton gin trash as supplemental fuel for a coal-fired power plant near Lubbock, Texas (Williams et al., 1982).

As of 1980, there were 231 active gins located within an 80-km radius of Lubbock. Approximately one million bales of stripper-harvested cotton are ginned annually. The number of bales ginned per year averages 4,300 per gin, with the volume for only two of the 231 gins exceeding 10,000 bales per year. The amount of trash available for use in a typical year would average 160×10^3 tonnes, or about 700 tonnes per gin. This study considered the feasibility of supplying 7 percent of the thermal input for a hypothetical 500-megawatt unit.

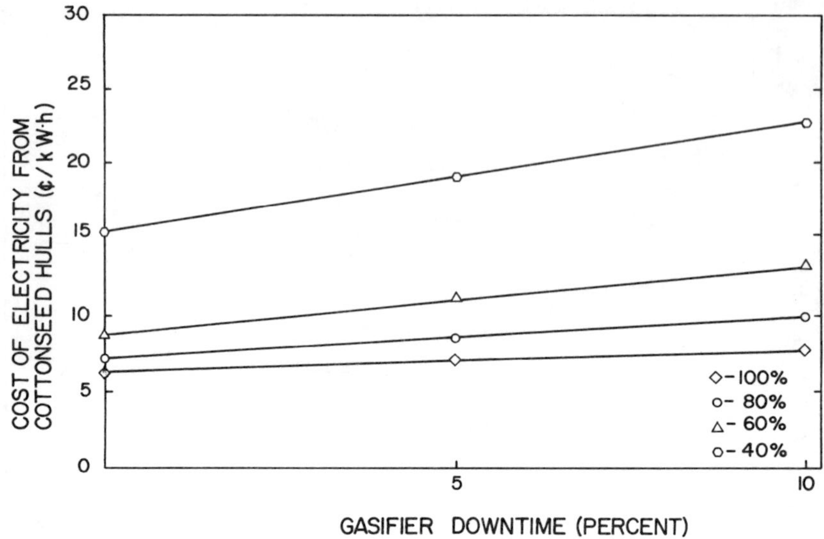

FIG. 6.8. Effect of percent downtime upon the cost of electricity generated as a function of percent mill utilization for a 200-tonne-per-day oil mill. Costs were based on power generating system life of 5 years, fuel costs of $10 per tonne.

The energy content of the coal used for power generation varies between 17.4 MJ and 25.6 MJ per kilogram. The current (1984) cost of coal varies between $30 and $50 per tonne.

In the systems analyses, the two major cost items were the costs associated with the acquisition and delivery of the gin trash to a plant location, and the materials-handling costs at the plant. Acquisition and delivery expenses included payments (to the gin) for the trash, moduling, storage, and transportation.

Loose gin trash is bulky, difficult to handle, and subject to wind and decomposition losses during storage. In these analyses, it was assumed that the trash would be formed into 9.1-tonne (10-ton) modules soon after ginning. (The moduling equipment is the same type that is currently used for moduling seed cotton.) It was also assumed that a module builder would be required exclusively for moduling gin trash.

The gin trash modules were transported to a storage area near the gin, where they would remain until needed by the power plant (see Figure 6.10). It was assumed that storage space could be leased, with space requirements averaging 125 modules per hectare. The storage costs were assumed to be based on an annual charge of $247/ha ($100/

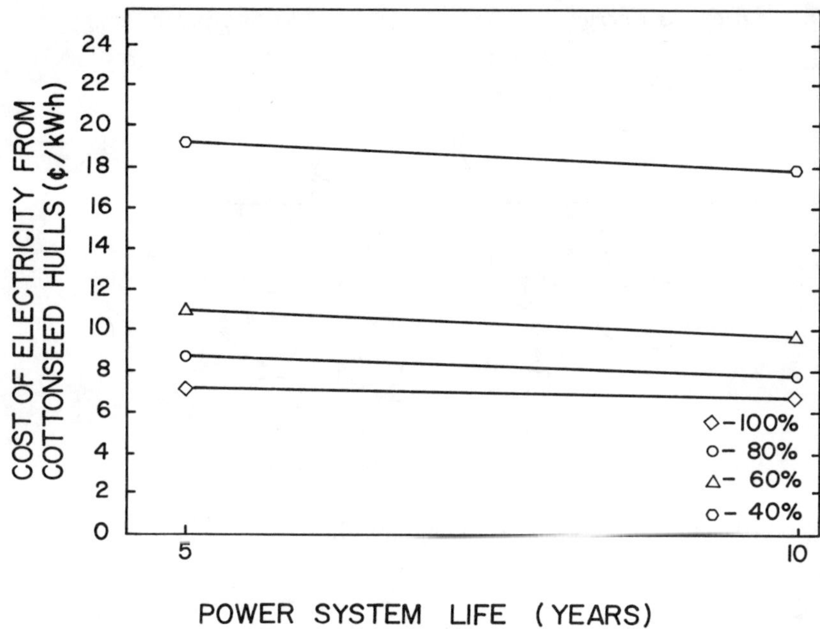

FIG. 6.9. Effect of power system life on cost of electricity generated as a function of percent mill utilization for a 200-tonne-per-day oil mill. Costs were based on power system generation downtime of 5%, fuel cost of $10 per tonne.

acre). In addition, a charge of $2 per module for transportation from the gin to a storage location was anticipated.

Estimated costs for transportation from storage to user location were based on rates published by the Texas Railroad Commission (1981). The distance for each of the 231 gins within the study area was determined. Total transportation costs for each gin were estimated by multiplying the volume of trash for each one by the published transportation rate for the determined distance.

In order for power plants to seriously pursue gin trash biomass as a possible fuel source, two conditions must be satisfied: (1) there must be economic incentive for the utility company, and (2) a large percentage of the cotton ginners must be willing to guarantee delivery of their gin trash at a specified price for several years in advance.

The fuel costs for a 500-MW coal-fired power plant are calculated in Table 6.6. For coal delivered at $30 to $50 per tonne, having an energy content of 17.4 to 25.6 MJ per kilogram, the energy costs

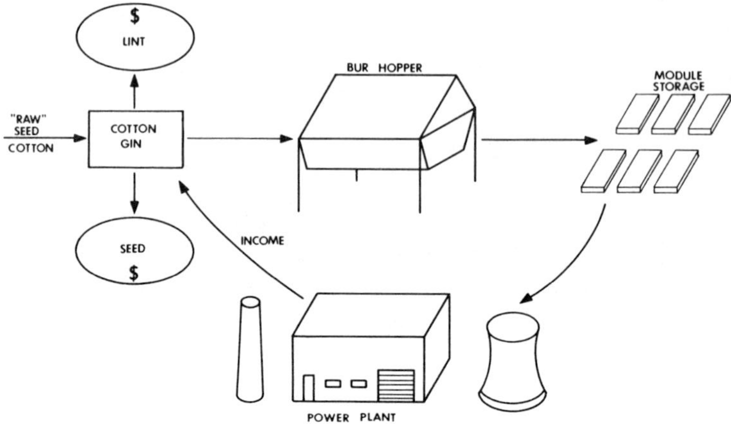

FIG. 6.10. Diagram of the power plant model.

would be $1.17 to $2.87 per gigajoule. For cotton gin trash delivered at $15 to $25 per tonne, having an energy content of 15.1 to 17.4 MJ per kilogram, the fuel costs would be $0.86 to $1.65 per gigajoule.

In addition to this cost, the power plant must install a materials-handling system completely separate from the coal system, and purchase the necessary module storage and handling equipment. It is estimated that this would add as much as $0.50 per gigajoule to the cost of using gin waste.

The objective of the simulation testing was to answer the following questions:

1. At what distance from the power plant would it be economically feasible for ginners to deliver gin trash?

2. At a fixed distance from the power plant, how much gin trash must be sold for ginners to break even?

3. How much does each individual cost contribute to the total cost of biomass collection, storage, and transportation to the power plant?

4. What would be the fixed and variable costs at the gin?

5. What benefits would be realized from this activity for (a) the ginner's goal, which is to process seed cotton at minimum cost, and (b) the power company's goal, which is to generate electricity at least cost?

The system under consideration was conceptualized as shown in Figure 6.11. The simulation model was composed of three major sub-

TABLE 6.6 Fuel Cost per Gigajoule for a 500-MW Power Plant (30% thermal conversion efficiency)

	Coal				Cotton Gin Trash			
Energy content	17.4 MJ/kg		25.6 MJ/kg		15.1 MJ/kg		17.4 MJ/kg	
Energy in	6.0 TJ		6.0 TJ		0.6 TJ		0.6 TJ	
Fuel in	344 tonne/h		235 tonne/h		40 tonne/h		34 tonne/h	
Fuel cost	$30/tonne	$50/tonne	$30/tonne	$50/tonne	$15/tonne	$25/tonne	$15/tonne	$25/tonne
Energy cost	$1.72/GJ	$2.87/GJ	$1.17/GJ	$1.95/GJ	$.99/GJ	$1.65/GJ	$0.86/GJ	$1.44/GJ

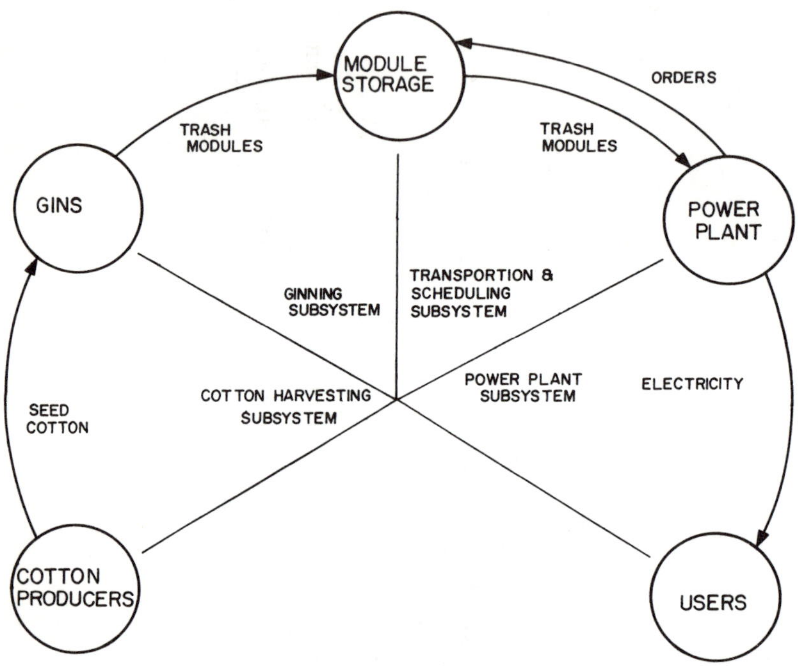

FIG. 6.11. Conceptualized gin trash utilization system.

systems: (1) ginning, (2) transportation, and (3) power station. Loads of seed cotton arrived at the gins and were processed. Trash generated by the ginning process was compressed into modules and stored. As trash was required, it was transported to the power station, where it was burned as fuel.

Gin Subsystem

The gin subsystem portrays all gins located in the power plant's vicinity. The number of gins and their distances from the power plant were specified through input data. Although each gin is technically a subsystem for this study the entire set of gins was considered "the gin subsystem." Gins were grouped according to their distances from the power plant. Trash from gins in the first (closest) group would be transported to the power plant first. After consuming this trash, the power plant would order from the next closest group. The process would continue until there was no trash left to supply the power plant, or the simulation stop-time had been reached.

The cotton ginning operations were considered a subsystem in the model and the subsystem boundary was defined as the physical confines of an actual ginning operation, including equipment within the gin itself and the storage area for seed cotton and modules of gin trash. Seed cotton was assumed to enter the system boundary at the input of the gin. Trash, cottonseed, and baled lint emerge on the output side. Since finished lint and cottonseed are not of interest in this study, only trash accumulations were considered in this model. It was assumed that two extra workers would be required during the season to handle the moduling and storage of gin trash; their cost was assumed to be minimum wage. All costs associated with trash handling and storage were considered the responsibility of the ginner.

Transportation Subsystem

This subsystem incorporates all the activities associated with trash transportation from gins to power plant. The proposed method of transportation was the cotton module truck. When an order for gin trash was received at the gin, a truck would be dispatched to transport the module. If the order came in after business hours, or if the truck could not return before the gin's normal quitting time, then the order would be processed on the next business day. Truck speed was assumed to be triangularly distributed with a minimum of 64 kph, a mode of 72 kph, and a maximum of 80 kph. Upon arrival at the power station unloading area, the truck would unload the module of trash. Only one unloading facility was assumed for the plant; hence, if another truck were occupying the unloading area, the incoming truck would wait in line. The unloading operation was assumed to take one minute. Then the truck would return to the gin. Hauling costs were specified by the Texas Railroad Commission (Figure 6.12).

Power Station Subsystem

The power station was assumed to use gin trash to supply 7 percent of its thermal input. This translated into approximately three modules of gin trash per hour.

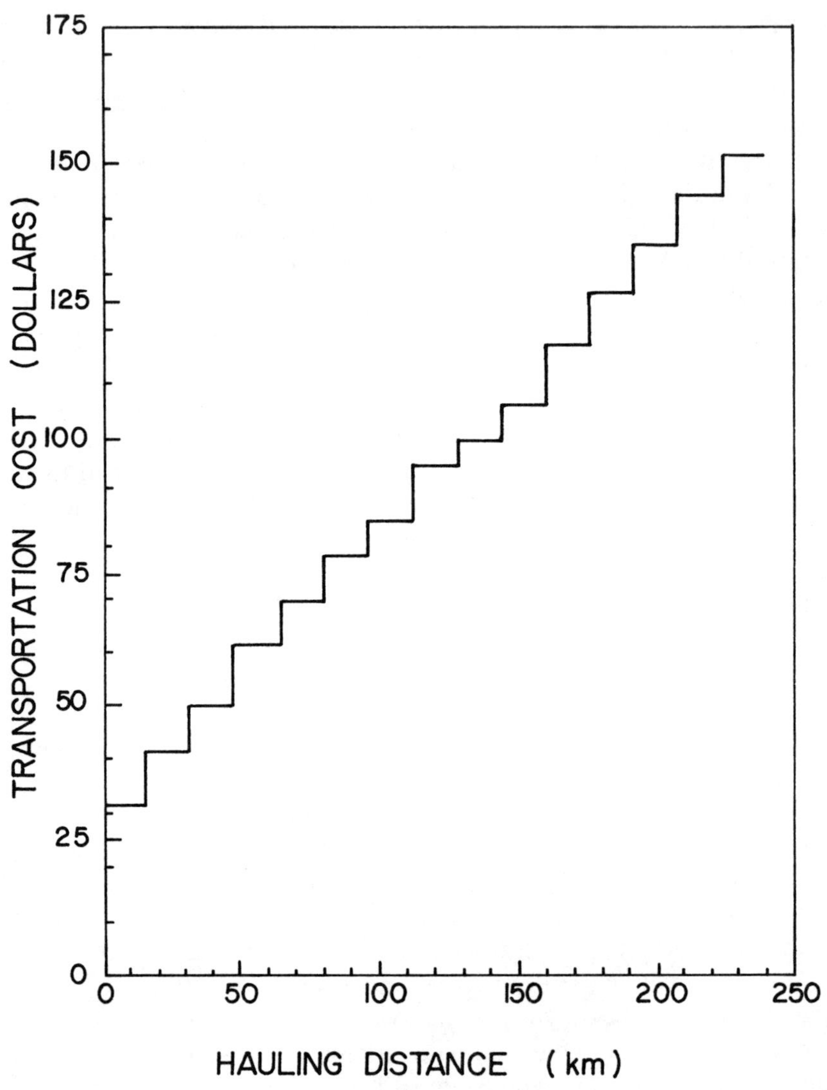

FIG. 6.12. Round trip hauling costs per 9.1 tonne module (from Texas Railroad Commission, 1981).

In order to supply enough gin trash to the power plant once it was being utilized (after the end of the ginning season), two quantities were monitored continuously. The first quantity was the on-hand, or simple, inventory. This variable reflected the amount of gin trash available for burning at the power station. Inventory on hand had no direct role in the decision of when to place an order for gin trash; it was monitored solely as an indicator of the performance of the inventory policy. The time profile plot of this quantity reflected any gin trash shortage at the power station and how long the shortage lasted.

The second quantity, the inventory position, reflected the sum of the on-hand inventory and the outstanding orders. When the inventory dropped below a threshold level (the reorder point), an order was issued for additional modules. The procurement procedure then caused a search for the closest gin that had at least one load of gin trash. The order was placed at that gin and the inventory procedure resumed its monitoring process. The reorder point effectively established a buffer stock of gin trash to hedge against the possibility of the power station's running out. The system was initialized with a ten-day supply of gin trash (720 modules) at the end of the ginning season. Ten days later, the power station began utilizing the gin trash. The reorder point was set at 300 modules (100 hours of operation).

Table 6.7 is a summary of the simulation results for a power plant operated 352 days fueled with coal and gin trash. The plant purchased 225,000 tonnes of gin trash, yielding a total income of $1,240,000 for the ginning community. In addition, $1,370,000 was paid to ginners for hauling this product to the power plant. The aggregate cost to the power plant was $11.60 per tonne.

The gins in this study operated for 40 days (normal gin season), processing 670,000 bales. The ginners received $4.54 per tonne ($5/ton) of trash. However, the costs of moduling, handling, and storage were such that the 231 gins averaged a net loss of $0.36 per tonne. When trash was delivered from beyond 60 km, a loss was incurred. If the gin did not market more than 934 tonnes of trash, which is equivalent to ginning 3,050 bales of cotton per season, it incurred a loss. The gins nearest the plant showed profits of up to $3.57 per tonne, while the gins farthest from the plant lost up to $4.25 per tonne. When the gins received $11 per tonne ($10/ton) for their trash, however, the average profit rose to $3.96 per tonne, and profits of up to $8.64 per

TABLE 6.7 Summarized Results of Power Plant Simulation

Power plant report	
Trash purchased	225,000 tonnes
Length of burn	352 days
Purchase cost	\$1,240,000
Hauling cost	\$1,370,000
Aggregate cost	\$11.60 per tonne
Average inventory	2,340 tonnes
Maximum inventory	6,530 tonnes
Gin report	
Average gin year	40 days (3,055 bales ginned)
Total seed cotton processed	670,000 bales
Price received for trash	\$4.54 per tonne
Economical delivery distance	60 km
Average break-even point for the gin	934 tonnes sold
Average net profit or loss (over 231 gins)	−\$0.36 per tonne

tonne were realized by the gins nearest the plant. Six of the 231 gins did not show a profit, but the maximum loss was only \$1.00 per tonne. Since it currently costs ginners between \$6.50 and \$10 per tonne to dispose of their gin trash, a net loss of \$0.36 per tonne could be viewed as a net profit of \$6.14 to \$9.64 per tonne.

This simulation model offers the opportunity to analyze a number of variables in the light of both the power plant's and cotton gin's objectives, and answer such questions as the following: (1) What quantity of biomass would be available within a specified radius of the user location? (2) At a consumption rate of 27.3 tonnes (three modules) of gin trash per hour, how long could a power plant expect to have this biomass as a supplemental fuel? (3) What would be the schedule for ordering gin trash, and were there any periods of time when gin trash was not available at the power plant?

Utilizing gin trash as a fuel supplement can help offset at least a portion of the cost of producing electricity and at the same time provide a more profitable means for disposal of gin waste materials.

Energy-Integrated Farm Systems

One application of systems engineering is the design and employment of energy-integrated farm systems (EIFS). EIFS demonstrations were started by the U.S. Department of Energy in September 1980 and are in process at this time. The goal of the program is to demonstrate new technology that will effectively reduce on-farm need for energy derived from fossil fuels. The objectives are "(1) to reduce to a minimum the external dependence of farms on scarce energy resources such as petroleum, and (2) to develop energy farms where a combination of existing or developing technologies surrounding renewable energy resources can be incorporated into a coordinated system with specific farm enterprises" (Olver, 1983). In reality, this program is an attempt by the DOE to speed up technology transfer. The concept of on-farm energy self-sufficiency can be traced back to animal agriculture (Hirning et al. 1982), where approximately one-third of the farm was devoted to producing feed for the animals that were then the primary source of energy.

Seven locations were chosen: five are private farms, and two are operated by universities. Table 6.8 provides a short description of each and the energy conversion technologies to be implemented. Note that anaerobic digestion seems to be very popular, indicating that this technology may be coming very close to economic feasibility.

The concept of demonstrating energy independence, or near independence, presupposes that the current level of technology has advanced to the point that success can be achieved. The complexity of using three or more biomass conversion or solar energy alternatives, each dependent upon another, decreases the probability of successfully developing alternative energies that are economically competitive with those derived from fossil fuels. However, success for the EIFS may be better defined as energy self-sufficiency for agriculture, irrespective of economics.

The view that biomass conversion technology will not be accepted and utilized unless economic incentives are evident is not necessarily true for on-farm energy self-sufficiency. Reliable energy resources are a necessity for many agricultural operations. For example, a crop could be lost if energy supplies were not available for planting, crop drying, or processing when they were needed.

Systems engineering of energy-integrated farm systems allows

TABLE 6.8 Summary of DOE-Sponsored Energy-Integrated Farm Systems, 1980–84

	Puerto Rico	Georgia	Nebraska	North Dakota	Texas	Virginia	New York
Location	Granja Caribe Poultry Farm and University Puerto Rico	Aubrey Farm and Georgia Institute of Technology	University of Nebraska Field Laboratory Farm	North Dakota State University Agricultural Experiment Station	Del Valle Hog Farm and SUMX Corporation	Foxlease Farm and Archbold Investment Company	Millbrook Farm and Cornell University
Project director	University of Puerto Rico, Rios Piedras, Puerto Rico 00931	Richard S. Combes, Georgia Institute of Technology, Atlanta, Georgia 30332	William E. Splinter, University of Nebraska, Department of Ag. Eng., Lincoln, Nebraska 68508	Harvey J. Hirning and James A. Lindley, North Dakota State University, Department of Ag. Eng., Fargo, North Dakota 58102	David A. Malish, SUMX Corporation, P.O. Box 14864, Austin, Texas 78761	John D. Archbold, Foxlease Farms and Archbold Investment Co., P.O. Box 2394, Upperville, Virginia 45357	L.P. Walker, Cornell University, Department of Ag. Eng., Ithaca, New York, 14850
Farm Characteristics	16-acre farm, 36,000 hens	600 dairy cattle, 250 milking cows	rotating corn, soybeans, and sweet sorghum	500 acres cropland, 100 milk cows, 980-sq.-ft. residence	2,990 acres grain	1074 acres cropland, 200 acres pasture, 200 acres woodland	455 acres total, 300 dairy cattle, 180 milking cows

Energy alternatives demonstrated	anaerobic digester; solar	alcohol still; anaerobic digester; solar; electricity generation from methane	anaerobic digester; solar; alcohol still	solar; anaerobic digester; alcohol still	anaerobic digester; electricity generation from methane; alcohol still	electricity generation from methane; solar; anaerobic digester; alcohol still	anaerobic digester; alcohol still; electricity generation from methane
Expected energy savings	NA	16,000 kWh/mo; 10,000 gallons of diesel/yr; 9,700 gallons gasoline/yr; 2,000 gallons propane/yr	NA	3,940 gallons fuel oil/yr; 5,000 kWh to residence	5,500 gallons diesel/yr; 7,000 gallons gasoline/yr; 47 kWh per day	5,900 gallons diesel/yr; 6,400 gallons gasoline/yr; 184,000 kWh/yr	438 barrels oil/year
Total cost	$1,013,000	$688,500	$1,720,400	$797,400	$818,000	$1,353,900	$1,532,000
Government share	$638,000	$378,700	$860,200	$398,700	$409,000	$677,000	$766,000
Private funds	$375,000	$309,800	$860,200	$398,700	$409,000	$676,900	$766,000

the designer to evaluate the effect of each energy alternative upon the system's performance prior to the installation of expensive hardware. For on-farm energy-integrated systems, reliable energy supplies during critical time periods for planting, harvesting, and processing should be emphasized. The major benefit of on-farm energy self-sufficiency is that it minimizes the effects of the price and supply of imported oil on U.S. agricultural production of food and fiber. An additional benefit is the transfer of biomass conversion and solar energy technologies while fossil fuels supplies are not critical.

Summary

The objective of a biomass-fueled energy alternative, a system that significantly reduces dependence on energy derived from fossil fuels, is technologically and economically feasible. The design of a biomass energy conversion system must incorporate the most recent technological advances. In addition, factors such as changing fuel costs, reliability, labor requirements, environmental regulations, waste disposal, and safety must be evaluated during the design phase and prior to the purchase of hardware.

In order to insure a high probability of success, a systems engineering approach to designing and analyzing the potential cost/benefit of a proposed system is recommended. The systems engineering approach is dynamic, with continuous feedback. An example of the steps that would be included in the design of a small power plant or cogeneration system is provided in Figure 6.2. This approach should incorporate (1) site-specific feasibility analyses, (2) simulation and testing, and (3) pilot-plant studies. The use of simulation techniques is emphasized.

Two examples of the power of simulation as a design tool are presented in this chapter. One considers the installation of a small power plant at a cottonseed oil mill; the second considers the use of cotton gin trash as a fuel extender at a coal-fired power plant. In both examples, the cost of implementing technology to acquire test data would be prohibitive. Simulation modelling and testing allows engineers to evaluate the effect of technical changes on the potential success of biomass conversion systems in the absence of actual test data. At this time, the simulation studies suggest that a small power plant at a large cottonseed oil mill fueled by cottonseed hulls or cotton gin trash might be attractive. The benefit of using gin trash to extend coal in a conventional coal-fired power plant is highly dependent upon coal costs and biomass supply.

The energy-integrated farm system demonstrations sponsored by the DOE utilize biomass conversion, solar, and conservation technology to displace energy derived from fossil fuels. The objective of these demonstrations is technology transfer; the concept is to demonstrate the potential for on-farm energy self-sufficiency, which may be needed in the future. For this reason, its justification may be based more upon reliability at critical times than on economy. The systems engineering approach to EIFS designs would be advantageous.

References

Allison, G. L. (1981). Co-generation and small power production. *Oil Mill Gazeteer* 86 (6): 31–34.

Baker, R. V. (1977). Cleaning machine stripped cotton for efficient ginning and maximum bale values. Agricultural Research Service Technical Bulletin 1540. Washington, D.C.: USDA.

Datin, D. L. (1983). Particulate cleanup of low-energy gas produced in a biomass fluidized-bed gasifier. Ph.D. dissertation, Texas A&M University, College Station, Tex.

Federal Energy Regulatory Commission (1978). Public Utility Regulatory Policies Act (PURPA). *Federal Register* 45 (83) 28162–64.

——— (1980). Rules by the Federal Energy Regulatory Commission. *Federal Register* 45 (25): 7819–21.

Gavett, E. E. (1980). Overview of energy consumption in U.S. agriculture—past and future. *Proceedings, American Society of Agricultural Engineers, National Energy Symposium*, 1:20–24. ASAE publication 3-81. St. Joseph, Mich.: American Society of Agricultural Engineers.

Good, N. E. (1981). Fuel from biomass. In R. A. Fazzolare and C. B. Smith, eds., *Beyond the energy crisis: Opportunity and challenge*, 2:491–98. New York: Pergamon Press.

Gustaferro, J. F. (1981). Energy for the rest of the century. In R. A. Fazzolare and C. B. Smith, eds., *Beyond the energy crisis: Opportunity and challenge*, 1:247–62. New York: Pergamon Press.

Hirning, H. J., J. A. Lindley, R. Johnson, J. Giles, G. L. Pratt, R. L. Witz, L. Roehl, and J. Fitts (1982). *Design for energy management on a northern great plains dairy/grain farm*. ASAE paper 82-3047. St. Joseph, Mich.: American Society of Agricultural Engineers.

LePori, W. A., C. B. Parnell, Jr., and P. Piumsomboon (1983). Biomass energy conversion to steam using fluidized-bed gasification technology. Paper presented at Solar and Biomass Workshop, Atlanta, Ga. Sponsored by the National Service Foundation, Washington, D.C.

National Academy of Sciences (1979). *Energy in transition 1985—2010.* Final report of the Committee on Nuclear and Alternative Energy Systems, National Research Council. San Francisco: W. H. Freeman.

Olver, E. F. (1983). Energy integrated farm systems program. Paper prepared for Energy from Agricultural Products and Residues Seminar, Austin, Tex.

Parnell, C. B., Jr. (1983). Systems analysis of cotton gin trash utilization alternatives. Vol. 1. Final Report Submitted to National Science Foundation. College Station, Tex.: Texas A&M University.

Parnell, C. B., Jr., W. A. LePori, and P. Piumsomboon (1984). Energy from cotton gin trash. In *Proceedings, 1984 Beltwide Cotton Production Research Conferences*. Memphis, Tenn.: National Cotton Council.

Pritsker, A. A. B., and C. D. Pegden (1979). *Introduction to simulation and SLAM*. New York: Wiley.

Public Utility Commission of Texas (1983a) *Section 210 of the Public Utility Regulatory Policies Act: A handbook for small power producers*. Austin, Tex.: PUC Economic Research Division.

——— (1983b). *Cogeneration and small power production in Texas: A survey and analysis of current regulatory issues*. Austin, Tex.: PUC Economic Research Division.

Strother, J. W. (1982). Moduling gin trash for cattle feed—a ginner's view. *Proceedings, cotton gin trash utilization and alternatives*. College Station, Tex.: Texas A&M University.

Texas Railroad Commission (1981). Hauling rates for unground cotton burs. Austin, Tex.: TRC.

U.S. Department of Energy (1980). Energy integrated farm systems. Publication No. DOE/TIC-11012. Washington, D.C.: Government Printing Office.

Williams, G. R., P. Piumsomboon, and C. B. Parnell (1982). A systems engineering model for cotton gin trash utilization. *Transactions of the ASAE* 25 (3): 802–805, 810.

CHAPTER 7

Economic Considerations

RONALD D. LACEWELL, SHARIF M. MASUD, RONALD C. GRIFFIN and GLENN S. COLLINS

Agricultural biomass is a possible alternative to some of the increasingly scarce and costly traditional energy sources. Although the production of substantial quantities of energy from agricultural biomass is technically feasible, considerable uncertainty exists over economic feasibility.

About 2 percent of U.S. energy use is supplied by wood, grasses, agricultural crops and residues, animal wastes, and other biomass sources. However, it has been estimated that from 5 to 17 EJ/yr could be produced from biomass by the year 2000 [Office of Technology Assessment, 1980a; Pimentel et al., 1983]. This represents 15 to 21 percent of the current U.S. energy use of 79 Quads/yr. These estimates are sensitive to national policy, food needs, and the cost of producing energy from biomass relative to alternative energy feedstocks.

In addressing the cost of producing energy from biomass, this chapter can be viewed as a combination of several case studies where alternative forms of biomass are used, different conversion technologies applied, and different types of organizations assumed. The first section discusses the use of on-farm cotton gin trash to produce a low-energy gas, as well as a large cooperative for the production of low-energy gas from gin trash and grain sorghum residues. The second examines both on-farm and gin-site electricity generation from gin trash. The third evalutes the anaerobic digestion of hog manures for biogas production. The fourth outlines the factors affecting the production of ethanol by fermentation of sorghum residues as well as corn and sorghum grain. The final section considers the production of oils from sunflower seed, cottonseed, peanuts, and soybeans for use as diesel fuel substitutes. Each discussion is based on and uses material from several published studies.

While various processes are being developed and improved for converting agricultural residues into energy, an economic analysis has not been completed on all options and systems. The potential biomass feedstocks considered here include cotton gin trash, grain sorghum residues, hog manures, feed grains, sunflower seed, cottonseed, and soybeans. For these systems and options, current economic estimates are useful in that they provide general implications and possibly point to the most economically feasible alternatives. Further, although a feedstock and conversion process may not be economically feasible at the current time, adjustments in supply and demand conditions could dramatically affect relative economic positions.

Much of the analysis presented indicates potential for the use of biomass as an alternative energy source. However, there are several limitations to many aspects of the technical and economic analysis, including the absence of commercially operating systems and the lack of experience in a commercial mode.

Gasification

The economic feasibility of producing a low-energy gas from agricultural residues assumed a fluidized-bed gasifier. The operation of a unit of appropriate size was evaluated for a farm, a large cooperative, and a cotton gin. Feedstock of gin trash and grain sorghum residues were evaluated.

On Farm

Cotton gin trash was considered as a feedstock for a fluidized-bed gasification system by Beck and Parker (1979). The analysis was based on a 640-acre (259.2-ha) irrigated farm that had four wells and good ground water (Lacewell et al., 1982). The region of study was the Texas High Plains, an area similar to the Great Plains reaching from Texas to Nebraska. Thus, the coefficients are applicable to those regions of the U.S. that use ground water for irrigation. The objective was to produce sufficient gas to exactly offset the irrigation energy requirements of a cotton farm (Table 7.1). It was assumed that the gin trash is put in modules of 9.1 tonnes (10 tons) each. The cost to purchase, module, transport, and store gin trash was taken from Moore, Lacewell, and Parnell, 1982.

The feedstock input rate was 544.3 kg/h. The system would be

TABLE 7.1 Energy Potential from On-Farm Gasification of Gin Trash to Offset Irrigation Fuel Use (640-acre cotton farm)

Item	Unit	Value
Energy in gin trash[(a)]	MJ/kg	16.3
Gin trash produced	1,000 kg	162.0
Total gin trash energy	TJ	2.6
Total energy recovery[(b)]	TJ	1.6
Energy used for irrigation	TJ	5.8
Additional gin trash required	1,000 kg	425.9
Days 4 wells pump	days	34.5
Days 2 wells pump	days	69.0
Fluidized-bed gasifier and automatic feeder	$	175,000.00
Fixed cost on investment	$/year	30,975.00
Fixed cost on investment	$/GJ	5.31
Operation cost[(c)]	$/year	12,322.00
Operation cost	$/GJ	2.11
Total cost[(d)]	$/GJ	7.42

Source: Masud, Lacewell, and Hiler (1983).

(a) Oursbourn et al. (1978).

(b) Beck and Parker (1979); 61% conversion.

(c) Derived from Beck and Parker (1979) by scaling down and deleting labor, income tax, and depreciation (part of fixed costs) and setting feedstock cost to $15 per ton (0.91 tonne) for moduling, handling, transportation, and storage (Moore, Lacewell, and Parnell, 1982).

(d) Does not include cost of distribution, gas cleanup, or derating an engine for lower-energy gas.

used at about one-tenth of capacity, since irrigation does not occur throughout the year (Masud, Lacewell, and Hiler, 1983).

For this system, operation costs alone (not including any fixed costs) approach current natural gas cost. Because the system would not be used all year long, cost is very high: $7.42 per GJ or $7.83 per million Btu (See Table 7.1). This is compared to a current natural gas price of near $3.32 per GJ or $3.50 per one million Btu (Masud, Lacewell, and Hiler, 1983).

Extending the use of the system to 292 days per year is 80 percent utilization (Table 7.2) and approximates electrical generating facilities (Pollard, 1982). During one year the plant required 3.9 Gg of trash, or gin trash from about 4,374 ha of cotton production. In this system, gas

TABLE 7.2 Year-Round Operation of On-Farm Fluidized-Bed Gasification Unit

Item	Unit	Value
Total gin trash required	Mg	3.9
Total energy produced	GJ	3.8
Investment	$	175,000.00
Fixed costs	$/year	30,975.00
Operation cost[(a)]	$/year	34,937.00
Feedstock cost[(b)]	$/year	63,822.00
Total annual cost	$/year	129,734.00
Cost of gas production[(c)]	$/GJ	3.37
Energy produced beyond irrigation	TJ	32.6

Source: Masud, Lacewell, and Hiler (1983).
Note: Based on 80% utilization or 292 days of operation annually; unit size is 8.9 GJ in and 5.5 GJ out per hour. The gas produced must be cleaned up for use in an internal combustion engine (Pollard, 1982).
(a) Derived by scaling down Beck and Parker (1979) estimates. Costs of operation and maintenance are not well developed.
(b) Feedstock costs are considered as cotton gin trash purchased, moduled, transported, and stored at $15/ton or 0.91 tonne (Moore, Lacewell, and Parnell, 1982).
(c) Does not include cost of gas cleanup, distribution system, or derating an engine for lower-energy gas.

production beyond irrigation needs would have to be sold for credit or used in other farm activities, but the cost is much more attractive: $3.37 per GJ or $3.56 per one million Btu (Masud, Lacewell, and Hiler, 1983). However, the cost estimate does not include a cleanup charge or the cost of derating an engine to use the low-energy gas. Also, even clean low-energy gas is unsuitable for pipeline distribution. Thus, a direct comparison of low-energy gas with natural gas is not really feasible. The production costs of low-energy gas are expected to be slightly higher when using grain sorghum residue because of lower heat content and the need for field baling and hauling (Lacewell et al., 1982).

Large Farmer Cooperative

To provide insight into the feasibility of organizing a farmer cooperative to produce gas, a 907-tonne-per-day (1,000 ton/d) gin trash facility and a 2,720 tonne-per-day (3,000 ton/d) grain sorghum residue facility were considered (Masud, Lacewell, and Hiler, 1983).

TABLE 7.3 Best Case Large-Scale Cooperative Gasification of Gin Trash (907 tonne/d, or 1,000 ton/d)

Item	Unit	Value
Associated cotton lint	Gg	220.7
Gin trash	Gg	364.0
Total gas produced[(a)]	PJ	2.5
Cost of gasification[(b)]	10^6 $/yr	14.57
Cost of produced gas[(c)]	$/GJ	5.54

Source: Masud, Lacewell, and Hiler (1983).
(a) Based on Beck and Parker (1979) conversion of 61%.
(b) Based on wheat straw estimates of Beck and Parker (1979) with a feedstock cost of $10 per ton (0.91 tonne) for moduling, transporting, and storing gin trash (Moore, Lacewell, and Parnell, 1982), and deletion of income tax since the cooperative would not be directly taxed by the Internal Revenue Service.
(c) Does not include cost of gas cleanup for use in internal combustion engine, distribution of gas, or derating an engine for use of low-energy gas.

Cotton gin trash is fed to a fluidized-bed at 907 tonne/d for gasification (Table 7.3). A plant this size requires the gin trash from nearly three-quarters of a million bales of cotton: that is, 29,100 modules of trash at 9.1 tonnes each. This represents the output of approximately 1,600 farmers with 200 ha of cotton each. The costs of production are estimated to be $5.54/GJ or $5.84 per million Btu—beyond economic feasibility, particularly since there is no consideration of gas cleanup or distribution (Masud, Lacewell, and Hiler, 1983).

Since grain sorghum residue may be much more dense than gin trash with the added biomass production, and Beck and Parker (1979) indicate some economies of size, a 2,720 tonne/d (3,000 ton/d) gasification facility was considered. Sorghum varieties are currently being developed that will produce in excess of 11.2 tonnes of residue per hectare (Miller, 1980). Assuming 11.2 tonnes of residue per ha with 2.2 tonnes remaining on the land, 9.0 tonnes will be available for energy production. Some residues cannot be harvested and 0.91 tonne per acre is required for erosion control and to protect soil productivity (Office of Technology Assessment, 1980). For a 2,720 tonne/d plant, grain sorghum residue from 88,300 ha of grain sorghum production would be required.

TABLE 7.4 Best Case Large-Scale Cooperative Gasification of Grain Sorghum Residue (2,720 tonne/d, or 3,000 ton/d)

Item	Unit	Value
Sorghum residue energy[(a)]	MJ/kg	13.9
Sorghum residue required	Gg	792.0
Sorghum residue required	1,000 ha	88.4
Total gas produced[(b)]	PJ	6.7
Cost of gasification[(c)]	10^6 $/yr	24.2
Cost of produced gas[(d)]	$/GJ	3.58

Source: Masud, Lacewell, and Hiler (1983).
(a) Oursbourn et al. (1978)
(b) Based on Beck and Parker (1979) conversion of 61%.
(c) Based on Beck and Parker (1979) estimates with $6.9 million income tax deleted. The feedstock cost of $10 per ton (0.91 tonne) approximate cost to bale and haul grain sorghum residues with 9.0 tonne per ha harvested. There is no charge allotted to purchasing feedstock.
(d) Does not include cost of gas cleanup or distribution, or any payment to farmer for feedstock, or any allowance for storage of a very large quantity of residue.

The quantity of gas that could be produced is 6.7 PJ, or enough to irrigate over 486,000 ha with a low-pressure sprinkler system. But at a cost of $3.58 per GJ, not including the costs for gas cleanup and derating an engine for the low-energy gas (Table 7.4), it can be concluded that this large cooperative gasification system is not currently economically feasible if natural gas is available at $3.32 per GJ (Masud, Lacewell, and Hiler, 1983).

Electrical Generation

This section considers the economic implications of an on-farm electrical generation unit and one located at a cotton gin. In both cases cotton gin trash is used as a feedstock.

On Farm

The on-farm electrical generation unit is assumed to include a fluidized-bed gasifier, a boiler, and a turbine. The size is sufficient to satisfy the energy requirements of four irrigation wells on a typical Great Plains farm. Surplus electrical generation would need to be put into a grid system for credit.

This electrical generation unit requires about 500 kg of gin trash per hour, or 3.5 Gg a year, based on 292 operating days. The total investment for the plant is $275,000; annual fixed costs would be about $50,000; the cost to module, deliver, and store cotton gin trash, $57,816; and operating expenses, $39,560. This implies an annual outlay of nearly $150,000 to produce 2.4 GWh of electricity, amounting to a cost of $0.615 per kWh (Table 7.5).

Some $8,900 of electricity purchases are avoided by using the on-farm generation plant, and two million kWh of electricity is available for sale. This study used a credit of $0.02 per kWh, the typical amount paid to small producers of electricity that feed power back to a grid and generally equal to the cost of electricity generation by the grid. Naturally, this credit (as well as the cost to consumers) for electricity can vary widely from region to region.

The net result of electrical generation on farm is $104,171 of costs not covered for the producer. Of course, adjustment in the cost of electricity or of the system could significantly affect these results (Masud, Lacewell, and Hiler, 1983).

Cotton Gin

The most advantageous place to use gin trash for energy is at the cotton gin (Lacewell, Taylor, and Hiler, 1981), where the trash is already centralized and available. For this analysis, a gin size of 40,000 bales (9,070 tonnes) per year was selected, which is much larger than the average cotton gin. This gin would have 13,600 tonnes of gin trash available, or 1,500 modules of 9.1 tonne each (Table 7.6).

It is assumed that only the gin trash available at this particular gin would be used for electricity generation. For a fluidized bed, the rate of use of gin trash would be 1.9 tonnes per hour, based on 292 days of operation per year (Masud, Lacewell, and Hiler, 1983).

For the basic situation outlined above, the implications of using a fluidized bed in conjunction with a boiler and turbine to generate electricity were investigated. The expected investment in equipment was $868,400; the electricity produced, 7.78 million kWh per year. The gin uses 2.4 million kWh per year of electricity for ginning and a like amount of energy for drying the cotton (Table 7.7).

Annual costs of generating the 7.78 million kWh of electricity is estimated to be $412,000. This implies about $0.053 per kWh, which is competitive with current costs of electricity in the region (Table 7.8).

TABLE 7.5 On-Farm Electrical Generation Using Cotton Gin Trash with Credit for Surplus

Item	Unit	Value
Electrical output of generator	kW	339
Equipment investment	$1,000	275
Other start-up cost[(a)]	$1,000	13.2
Cotton gin trash required[(b)]	1,000 kg	3,496.7
Energy produced	TJ/year	56.6
Energy produced	1,000 kWh	2,375.7
Energy not used for irrigation	1,000 kWh	2,094.0
Fixed cost on investment	$/year	48,675.00
Feedstock cost	$/year	57,816.00
Operating cost[(a)]	$/year	39,560.00
Total cost[(c)]	$/year	146,051.00
Total cost[(c)]	$/kWh	0.0615
Credit for sale	$/kWh	0.02
Total credit[(d)]	$/year	41,880.00
Net producer cost	$/year	104,171.00

Source: Masud, Lacewell, and Hiler (1983).
Note: Based on a 259.2-ha farm with four irrigation wells; unit size is 8.1 GJ in and 339 kW out per hour. It is assumed the system is operated year round using a fluidized-bed gasifier, boiler, and prime mover to generate electricity.
(a) Derived from Beck and Parker (1979) estimates and scaled down.
(b) This is trash from over 2,268 tonnes of cotton lint, or 385 modules.
(c) Does not include cost of distribution and assumes cotton gin trash is purchased, moduled, transported, and stored for $15 per ton, or 0.91 tonne (Moore, Lacewell, and Parnell, 1982). Operation and maintenance costs are not well developed.
(d) Includes a credit of $8,982 for irrigation fuel costs provided by the system.

However, the economic feasibility of electricity generation is not totally based on local rates for electricity. The case study estimates that the gin will use waste heat for drying. Also, all electricity requirements of the cotton gin are satisfied. The electricity and drying costs represent $204,000 of expenses not incurred by the gin. The 5.375 million kWh of surplus electricity (above gin needs) are sold to a distributor for $0.02 per kWh, which is far below the $0.053 cost to generate. Again, the $0.02 per kWh is a basic credit or buy-back rate currently practical in the region and amounts to the cost of fuel for electrical generation by the power company. These sales give an in-

TABLE 7.6 Characteristics of Gin Trash Utilization for Energy Production at the Gin Site

Item	Unit	Value
Gin size	Mg lint/year	9.1
Gin trash	Mg	13.6
Energy generation	days/year	292
Gin trash use	kg/h	1,941.8
Energy in trash	GJ/h	31.6
Fluidized-bed combustor	$ investment	520,000
Boiler and prime mover	$ investment	348,400
Gin energy use	kWh/bale	120

Source: Masud, Lacewell, and Hiler (1983).

TABLE 7.7 Economic Implications of Electricity Generation Using Gin Trash at the Gin Site

Item	Unit	Value
Energy input	GJ/h	31.6
Energy output	kW/h	1,321
Investment	$	868,400
Annual fixed cost	$	153,700
Cost of gin trash[(a)]	$/year	75,000
Variable costs of operation[(b)]	$/year	183,550
Total cost	$/year	412,250
Total energy produced[(c)]	MWh	7.8
Cost per unit	$/kWh	0.053
Gin electricity replaced	MWh	2.4
Electricity sold[(d)]	MWh	5.4

Source: Masud, Lacewell, and Hiler (1983).

Note: Based on a gin size of 40,000 bales (9,072 tonnes) per year; trash from this gin only; 31.6 GJ/h of gin trash burned in a fluidized bed with 17% efficiency in the electricity generation process from gas produced.

(a) Based on a gin trash cost of $5 per ton (0.91 tonne) to module and store with no transportation or purchase costs (Moore, Lacewell, and Parnell, 1982).

(b) Estimate based on Beck and Parker (1979). Cost of operation of the boiler and turbine are not well established.

(c) It is assumed that waste heat is used for drying the cotton.

(d) It is assumed that the buy-back rate for electricity placed into a grid system is $0.02/kWh. No provisions for distribution have been included.

TABLE 7.8 Summary of Electrical Generation at Gin Site Using Gin Trash

Item	Unit	Value
Energy produced	MWh	7.8
Surplus energy produced	MWh	5.4
Cost of generation	$/y	412,250
Cost of generation	$/kWh	0.053
Typical gin electricity cost	$/yr	168,000
Typical gin drying cost[(a)]	$/yr	36,000
Gin credits from generation		
Energy costs not purchased	$/yr	204,000
Electricity sold[(b)]	$/yr	107,500
Total		311,500
Net cost of electrical generation	$/yr	100,750

Source: Masud, Lacewell, and Hiler (1983).
(a) Given a gin size of 40,000 bales (9,070 tonnes) per year, which is much larger than the average.
(b) Based on 5.4 GWh sold at $0.02/kWh.

come of $107,500. Thus, the total economic benefits of electrical generation are an estimated $311,500. Since total costs of production are over $412,000, there are expenses of about $100,500 not covered (Masud, Lacewell, and Hiler, 1983).

Anaerobic Digestion of Animal Manure

Anaerobic digestion is the most widely used technology for converting animal manures to produce biogas and is also used as a resource management tool for animal confinement operations. Chapter 3 discusses the technology and its characteristics. An economic and financial analysis of an anaerobic digestion system must consider cash flow components, the investment and energy tax credits effect, interest and inflation factors, and detailed data on the system's investment, operation, maintenance, and output.

Overview

Smith et al. (1979) showed that most of the full-scale anaerobic digestion units reported in the literature focus on the operational, engineering, and biochemical aspects rather than the economics of the systems. A review of the literature indicates that since the technology is in a very early and evolving stage of development, the current state of the art does not favor investment of capital based on gas production alone. In almost all cases, the outlook on economic payback has been influenced by the inclusion of such useful by-products as effluent feed (or refeed credit) and fertilizer enhancement through nitrogen conversion and conservation, or by such induced needs as pollution abatement and odor control.

In an economic analysis, the Office of Technology Assessment (1980b) presented the general trend in economic feasibility of anaerobic digestion of animal manures for individual farms in the major producing states. It reported that a substantial quantity of the manure produced in livestock operations could be used economically to produce energy if the effective capital charges are reduced through various economic incentives.

The OTA analysis further reported that the principal cost factor in anaerobic digestion is the capital cost of the digester. Financing aside, anaerobic digestion systems are most economically attractive for relatively large poultry, dairy, beef, or swine operations that are also relatively energy intensive. Using 1978 fuel prices, the study assessed the feasibility of anaerobic digestion for various types of farm animal operations in the various regions of the country; concluded that it would be feasible to digest 50 percent of the animal manure to produce electricity and on-site heat if the effective annual capital charges were 6.6 percent of the investment. This effective capital charge could be achieved by a 20-year loan at 9 percent interest, with 4.2 percent annual tax writeoff (OTA, 1980b).

Hog Farm Anaerobic Digester

The purpose of this section is to examine a specific study in which an anaerobic digester is linked to a swine operation as carried out by SUMX Corporation on the Del Valle hog farm under contracts with the U.S. Department of Energy and the Texas Energy and Natural Resources Advisory Council. This is one of the energy-integrated farms referred to in Chapter 6.

The economic and financial analysis of the Del Valle hog farm's anaerobic digester is based on a study by Lacewell and Penson (1982). The digester constructed on the farm is a 113.2-m^3 (4,000-ft^3) unit.

The capital cost, operation and maintenance costs, and the service life and salvage value of each component of the unit are presented in Table 7.9. The overall investment is $61,738, with annual operation and maintenance costs of $3,240. Component service life ranged from 5 to 40 years, and the salvage value of the different components varied greatly.

Output Value. A major return to the digester operator is a reduction in the cost of purchased heating. For this analysis, fuels used for heating were considered to be propane and electricity. The net energy available for use is assumed to be 85 percent of the total energy produced, with 15 percent required for reactor heating. This assumption is based on average meteorological conditions at Austin, Texas (National Climatic Center). For the 113.2-m^3 digester, available biogas is 17,546 m^3 per year, or 392 GJ. Table 7.10 indicates reduced heating costs attributable to the digester. Assuming that only the biogas produced from October 1 to June 1 will be used for heating, only two-thirds of the total constitutes a heating credit (reduction in traditional fuel purchased). However, it was assumed that one-half the gas produced from June 1 to October 1 could be used for some alternative purpose, and its value as a natural gas replacement was given a credit of $500. The last column of Table 7.10 includes this $500 summer credit for biogas produced.

The total annual value of the biogas is then $2,500 if replacing natural gas, $4,500 if replacing propane, and $9,292 if replacing electricity (Lacewell and Penson, 1982).

Hog manure is rich in nitrogen. For this analysis it is assumed that of the 7.3 tonnes of nitrogen produced, one-half is lost in the digestion process, leaving 3.6 tonnes available to be applied to the sorghum acres included in the farm operation. A price of $0.44 per kg ($0.20 per pound) gives a credit of $1,600 annually to the nitrogen.

SUMX Corporation estimated that had the Del Valle hog farm not installed the anaerobic digestion facility, its operators would have incurred a $3,000 capital cost to upgrade the existing manure disposal unit. Since this investment was not made, there is a credit. Assuming that the upgraded equipment would last 20 years and have a zero salvage value, the $3,000 can be expressed as an annual credit of $480; i.e., $480 per year over 20 years is comparable to a total current value of $3,000. If the period were reduced to 10 years, the credit would be approximately $600 per year (Lacewell and Penson, 1982).

TABLE 7.9 Investment, Operation and Maintenance, Service Life, and Salvage Value for Digester Components

Component	Capital Investment ($)	Operation & Maintenance ($/year)	Service Life (years)	Salvage[(a)] Value ($)
Sump	4,050		40	0
Scraper	4,000	200	5	1,000
Reactor	14,700		40	0
Reactor top	6,000	50	20	0
Heating	1,920	214	5	150
Mixing	2,930	264	5	1,000
Liquid discharge	2,509		30	0
Solids removal	582		30	0
Gas cleaning	514	178	20	0
Intermediate storage	5,000		20	500
Pressure storage	5,000		40	4,000
Compressor	1,600	874	5	800
Operator		1,460		
Subtotal	48,805	3,240		
Site work & piping	7,321			
Total construction	56,126			
Engineering, legal, etc. (10%)	5,612			
Total project cost	61,738			

Source: Lacewell and Penson (1982).
Note: SUMX Corporation's Del Valle facility: 4,000 ft^3 (113.2 m^3).
(a) Represented in 1982 dollars. For this analysis, these values were calculated in 1992 dollars, using the expected annual rate of change in the prices of new farm machinery.

Land Cost. Supplying the anaerobic digester requires about 0.4 ha (1 acre) of land. This land was in coastal Bermuda grass. Thus, the annual value of production from the land is a cost to the facility. Typically, 70 bales (1.9 tonnes) of hay are produced on the 0.4 ha of land. It is assumed the hay would sell for $2.50 per bale ($92 per tonne) in 1982 dollars but cost $0.75 per bale ($28 per tonne) to cut and bale. Thus, the annual land cost is set at $122.50 (Lacewell and Penson, 1982).

Farm Financial Scenarios. Considering an investment tax credit of 10 percent makes the income tax situation of the potential investor a

TABLE 7.10 Heating Cost Reduction Attributable to Del Valle Anaerobic Digester

Fuel	Total Amount[(a)]	Amount Applicable[(b)]	Value ($)
Natural gas	21,255 m^3	14,150 m^3	2,500[(c)]
Propane	35,484.4 liter	23,656.3 liter	4,500[(d)]
Electricity	2.2×10^5 kWh	1.4×10^5 kWh	9,292[(e)]

Source: SUMX Corporation (1982).

(a) 392 GJ of biogas converted to natural gas, propane, and electricity heating equivalents.

(b) Only the biogas generated from October 1 to June 1 is needed for heating; i.e., two-thirds of the total can be used.

(c) Assumes a natural gas price of $141.3428 per 1,000 m^3 ($4 per mcf) and includes a credit of $500 for gas produced between June 1 and September 31.

(d) Assumes a propane price of $0.1717 per liter ($0.65 per gallon) and the $500 credit stated in (c) above.

(e) Assumes an electricity price of $0.06 per kWh and the $500 credit stated in (c) above.

critical component of the analysis. To provide some indication of the sensitivity of an economic analysis to tax liability, annual initial or base income tax liabilities of $6,000 and $12,000 were analyzed. The effect of the anaerobic digester on tax liability and overall cash flow was evaluated over a ten-year period.

The anaerobic digester was evaluated for a farmer who had assets of $649,756, debts of $228,804, and hence an equity situation of $420,952. This represents average conditions for an irrigation farm in the Great Plains (Federal Land Bank, 1980). For this study the farmer was assumed to be limited to a two-to-one ratio of debt to equity when borrowing to finance the purchase of the equipment as well as covering subsequent annual cash flow deficits over the life of the investment.

The term of the loan was set at five years. With an initial cost of $61,738, a 65 percent loan ($40,130) was assumed. The inflation and interest rates used in the analysis are presented by category in Tables 7.11 and 7.12. Two sets of interest rates for debt and equity capital were selected. The first set (Prime Interest Rate) was based upon the projected prime interest rate for debt capital and the prime minus three points for equity capital. The other set of interest rates for debt and equity capital was tied to projections for annual changes in the Consumer Price Index (Lacewell and Penson, 1982). For debt capital, this

TABLE 7.11 Inflation and Interest Rates Used for Economic Analysis of an Anaerobic Digester

	Basis for Rates	
Year	Prime Interest Rate (%)	Consumer Price Index (%)
Cost of equity capital[a]		
1	9.72	7.00
2	9.72	6.30
3	7.83	5.70
4	7.35	6.00
5	6.56	5.77
6	6.16	5.45
7	5.61	5.21
8	5.45	4.90
9	4.98	4.74
10	4.98	5.37
Cost of debt capital[b]		
1	15.30	11.90
2	15.30	11.00
3	12.70	10.20
4	12.30	10.70
5	11.30	10.30
6	10.80	9.90
7	10.10	9.60
8	9.90	9.20
9	9.30	9.90
10	9.30	9.80

Source: Lacewell and Penson (1982).
(a) Reported on an after-tax basis.
(b) Reported on a before-tax basis.

is the interest rate paid (interest on borrowed money). For equity capital (investment capital or money owned by the builder), this is interest forgone by building the digester.

Results. The economic analysis of an anaerobic digester is concerned with a replacement for propane and electricity for the heating of a swine operation. Initial $6,000 and $12,000 income tax liabilities un-

TABLE 7.12 Rates of Price Changes for Selected Cost Items Used in Economic Analysis of an Anaerobic Digester

Year	Prices Paid for Materials for Repairs and Maintenance	Prices Paid for Labor for Repairs and Maintenance	Annual Adjustments for Energy-Related Savings	Annual Adjustments for Other Cost Savings
1	7.7	8.9	4.0	7.5
2	5.9	8.0	7.9	8.2
3	4.9	7.2	13.3	10.0
4	4.8	7.6	14.1	13.9
5	5.1	7.3	12.1	14.2
6	5.2	6.9	7.0	13.0
7	5.1	6.6	7.3	11.3
8	5.0	6.2	8.6	10.3
9	4.6	6.0	9.0	9.7
10	4.4	6.8	8.1	10.2

Source: Lacewell and Penson (1982).

der two inflation interest rate scenarios—prime and CPI rates—were assumed. As a replacement for natural gas, the digester is clearly not economically feasible.

Assuming that the fuel used for heating is propane, Table 7.13 presents the results for a ten-year analysis (the values are discounted to 1982 dollars in all cases). The net present value is negative in all cases for propane, ranging from negative \$54,048 to negative \$42,203. Thus, even with favorable financing within the CPI scenario, the investment is not economically feasible. Financial feasibility is satisfied in all cases because of the initial high equity-to-debt ratio of the assumed investor.

When the reduced cost of an alternative fuel or income to an investment is not sufficient to satisfy annual costs for operation and maintenance plus interest and principal on the investment, a short-term cash flow loan is required. For propane, the short-term cash flow loan reached a maximum in years 5 and 6 at between \$26,281 and \$33,215. The anaerobic digester is not economically feasible as an alternative to propane unless there is an annual benefit for odor abatement and manure disposal of about \$5,000 or more per year (Lacewell and Penson, 1982).

TABLE 7.13 Financial and Economic Implications of a 4,000-Cubic-Foot (113.2 m^3) Digester Producing Methane from Pig Manure

Item	$6,000 Tax Liability[a]				$12,0000 Tax Liability[a]			
	Propane		Electricity		Propane		Electricity	
	Prime[b]	CPI[b]	Prime[b]	CPI[b]	Prime[b]	CPI[b]	Prime[b]	CPI[b]
Net present value ($)[c]	−51,722	−42,203	−6,213	2,743	−54,048	−49,197	−9,126	−3,546
Net cash flow ($)	−20,009	−13,033	17,355	24,198	−20,373	−16,342	13,645	17,814
Resale value ($)	9,623	10,754	9,623	10,754	−9,623	10,754	9,623	10,754
Remaining debt ($)	12,736	7,702	0	13,854	11,044	0	0	0
Equity	28,606	32,221	33,191	32,209	29,443	32,565	32,394	32,114
Cash flow loan[d]								
Maximum ($)	33,215	26,281	12,405	7,858	32,008	26,755	9,941	7,091
Year	6	5	5	5	6	5	5	5
Feasibility								
Cash flow	Yes	Yes	Yes	Yes	Yes	Yes	Yes	Yes
Economic	No	No	No	Yes	No	No	No	No

Source: Lacewell and Penson (1982).

(a) All situations assume a five-year loan for 65% ($40,130) of total initial cost ($61,738).

(b) Inflation and interest values for each year are presented in Tables 7.11 and 7.12.

(c) Net economic effect in 1982 dollars over a ten-year planning horizon. Economic feasibility requires net present value to be zero or greater.

(d) When sufficient savings in costs or returns to investment do not cover cost, a short-term cash flow loan is required. The maximum amount a cash flow loan reaches and the corresponding year are shown here.

The result of the anaerobic digester as a replacement for electricity in a pig operation is also presented in Table 7.13. With favorable financing as given by the CPI scenario and a $6,000 tax liability, the digester is economically feasible, giving a net present value of positive $2,743 over the ten-year evaluation period. Using the prime rate or a $12,000 tax liability leads to results that are not economically feasible, ranging from negative $3,546 to negative $9,126.

The cash flow loan reached a maximum in the fifth year in all cases. The maximum value ranged from $7,091 to $12,405. For the electricity replacement example, the $12,000 tax liability detracted from economic feasibility because any income advantages of the anaerobic digester were subject to the relatively higher tax rate, thus reducing the net present value of the investment as compared with a $6,000 tax liability. Nevertheless, the digester does indicate economic feasibility with favorable financing (CPI scenario) and an initial $6,000 tax liability (Lacewell and Penson, 1982).

Ethanol Production

Ethanol (alcohol made from grain or sugar crops) production for gasohol (one part ethanol and nine parts gasoline) has been an extremely popular proposal. To encourage ethanol alcohol production, the federal government has exempted gasohol from the federal excise tax. This exemption amounts to $0.04/gal ($0.01/L) of gasohol or $0.40/gal ($0.11/L) of ethanol. To further encourage gasohol sales, many states have exempted gasohol from the state road tax. This section summarizes and reviews some recent studies on the economics of on-farm ethanol production from grain and from sorghum residues.

Grain Feedstock

Several studies dealing with the economics of small-scale anhydrous ethanol production have been published recently and are reviewed by Bowker and Griffin (1983). The ethanol production plants considered in these studies have an annual capacity ranging from 3,250 to 1,000,000 L (860 to 264,000 gal) of 160- to 190-proof ethanol.

Overview of the Literature. The studies conducted by Atwood and Fischer (1980) and by Nichols and Jackson (1981) used the "economic engineering" approach that involves a "synthesis" of a particular operation on paper. Cost estimation data (obtained from suppliers and plant designs) and technical specifications (obtained from manufacturers) are incorporated into an evaluation framework that allows investment and operating tests to be estimated (Bowker and Griffin, 1983).

The studies done by Dobbs, Hoffman, and Lundeen (1981) and Growmark, Inc. (Farmer Cooperatives, 1981) used data obtained from a model plant. The type of plant involved can often be expanded by increasing its fermentation capacity without exceeding the initial distillation capacity. This allows the estimation of production costs for operations that are considerably larger than the model plant itself. For example, Dobbs, Hoffman, and Lundeen (1981) refer to the theoretical annual output of a specific facility as approximately 170,000 L per year; however, with increased fermentation capacity the same plant is capable of producing over 600,000 L per year (Bowker and Griffin, 1983).

Although the methods and intermediate results in the cited studies differ somewhat, the end results are quite consistent. Production costs for plants in the category of 28,000 to 45,000 L (7,500 to 12,000 gal) per year range from about $1.24/L ($4.75/gal) to $1.03/L ($3.90/gal) for 160- to 190-proof ethanol. Nichols and Jackson (1981) estimate $0.63/L ($2.37/gal) for the same product from a plant producing 760,000 L (201,000 gal) per year.

Regarding economic feasibility, the studies are again consistent. Ethanol production from corn and other feed grains will not be a profitable enterprise for the farmer. Atwood and Fischer (1980) reach a similar conclusion. Growmark, Inc. concludes that, "given present technology, small alcohol operations on individual farms are not practical because of high costs, handling problems with stillage, the nature of automobile and truck engines, and inefficiencies due to size limitations" (Farmer Cooperatives, 1981). Dobbs, Hoffman, and Lundeen (1981) claim that near-term economic prospects for nonsubsidized small-scale fuel ethanol production from corn do not appear very bright. They point out the economic problems of small-scale plants that must compete with much larger facilities producing anhydrous ethanol (Bowker and Griffin, 1983). A number of other studies with good documentation of investment requirements and costs have considered farm-sized plants with an annual production of 23,000 to 150,000 L (6,000 to 40,000 gal) of ethanol. Capital investment and cost estimates for three farm-sized plants with three types of operations, as discussed by Lacewell et al. (1982), are presented below.

The first is a farm-sized plant capable of producing 23,000 L (6,000 gal) of 180-proof ethanol per year, operating only six months (Fischer, 1980). The total capital investment of $73,500 was composed of $22,000 for a building to house the equipment and $51,500 for components, materials, and assembly. The capital investment per liter of annual capacity is $1.62 ($6.12/gal), assuming the full-time annual capacity of 45,000 L (12,000 gal) per year (Lacewell et al., 1982).

The total net production cost of $1.17/L ($4.42/gal) of ethanol makes this case uneconomical even with a large subsidy. This high cost results from two factors: (1) the large investment per liter of capacity, and (2) the part-time operation that increases the fixed cost per liter.

The second plant is capable of producing 150,000 L (40,000 gal) of 180-proof ethanol, operating year round (Fischer, 1980). The total capital investment for this plant is $104,000, and the capital investment per liter of annual capacity is $0.69 ($2.90/gal). The cost of production is $0.63/L ($2.40/gal). The fixed costs per liter are substantially lower and the variable costs per liter essentially the same, compared with the first case analyzed.

The third plant is a continuous-flow, semiautomatic plant designed to produce 760 L (200 gal) of 190-proof ethanol per day of operation, or 25,000 L (66,000 gal) per year (Butler Research and Engineering Co., 1980). This type of plant minimizes labor replacements and is considered to be more efficient than the batch-type plants. The estimated capital investment is $80,000, which is about $0.32/L ($1.21/gal) of annual capacity. The total net cost of $0.43/L ($1.62/gal), based on a price of $3 per bushel (25.4 kg) for corn already ground and available at the plant, is the lowest when compared with $1.17/L and $0.63/L, respectively, for the first and second cases analyzed (Lacewell et al., 1982).

The results indicate that 180- to 190-proof ethanol is unlikely to be produced by an on-farm plant for less than $0.42 to $0.53 per liter when all costs are included. Production costs for farm-sized plants are high because of the high investment relative to capacity and the resulting high fixed costs per liter. Investment cost per liter may be lowered in the future if improved small-scale technology can be developed. Production costs would also be lowered if (1) grain prices were lower, (2) by-product prices were higher, or (3) the conversion ratio were better than the 9 to 9.5 L/kg (2.4 to 2.5 gal/bu) used previously.

Small-Scale Plant. The capital investment requirements and costs of producing ethanol will vary by plant size. Two alternative plant sizes considered by Bowker and Griffin (1983) consist of an eight-hour-per-day operation of a plant producing 34,000 L (9,000 gal) per year and a 24-hour-per-day operation of one producing 545,000 L (144,000 gal) per year. The costs of producing ethanol can be separated into fixed and variable costs. The major items of the fixed costs include physical plant, taxes, insurance, and permits. The total fixed costs for the two physical plants are, respectively, $101,450 and $159,200 (Table 7.14). Over 12 years at a discount rate of 10 percent, the smaller plant represents an annual fixed cost of $14,889, or $0.44/L (1.66/gal) of ethanol produced. With the same time horizon and discount rate, the larger plant has an annual fixed cost of $23,365 or $0.05/L ($0.15/gal) of ethanol produced (Table 7.15). When the annual costs of physical plants are added to the annual costs of bonds and permits as well as insurance and property tax (Table 7.15), the results indicate that the smaller facility has a total annual fixed cost of $17,079 or $0.51/L ($1.90/gal) of ethanol produced, and the larger facility a total annual fixed cost of $26.973 or $0.05/L ($0.19/gal) of ethanol produced (Bowker and Griffin, 1983).

The major variable costs incurred in the distillation process, as discussed by Bowker and Griffin, include the cost of substrate (feedstock, including grain, sweet sorghum, and the like), labor, utilities (water, electricity, and natural gas), chemicals (yeast and enzymes), and maintenance when sorghum is used as a feedstock. The 34,000-liters-per-year plant has a total annual variable cost of $28,260 or $0.83/L ($3.14/gal) of ethanol produced; the 545,000-liters-per-year operation, $254,880 or $0.48/L ($1.77/gal) of ethanol produced (Table 7.15). These costs, however, are reduced because of the sale of the wet grain by-product. For both plants this by-product credit is estimated to be $0.06/L ($0.22/gal) of ethanol produced (Bowker and Griffin, 1983).

The total costs for both plants, summarized in Table 7.15, indicate that there are significant economies of size for the larger plant. While its annual cost is $250,173, as opposed to $43,359 for the smaller plant, the per-liter cost of ethanol is $0.47 ($1.74/gal) for the larger plant and $1.28 ($4.82/gal) for the smaller plant. Ninety percent of the difference between per-liter costs for these two plants is accounted for by better labor and plant utilization, which is due both to the increased capacity of the larger plant and to its continuous operation (Bowker and Griffin, 1983).

Insurance and property taxes for the larger plant are estimated to

TABLE 7.14 Physical Plant Fixed Costs

Item	Plant Size	
	34,000 L/yr	545,000 L/yr
Distillation equipment package	$ 87,450	$133,240
Feedstock storage bin	1,400	4,200
Assembly labor	800	960
Building	9,000	18,000
Dewatering press	2,800	2,800
Total	$101,450	$159,200

Source: Bowker and Griffin (1983).

TABLE 7.15 Cost Summary for Ethanol Production from Grain Sorghum

Item	Plant Size			
	34,000 L/yr		545,000 L/yr	
	Annual	Per L	Annual	Per L
Fixed cost				
Plant/facilities	$14,889	$0.44	23,365	$0.04
Bonds/permits	160	0.01	424	.00
Insurance/taxes	2,030	0.06	3,184	.01
Total	$17,079	$0.51	26,973	$0.05
Variable cost				
Sorghum substrate	$ 9,630	$0.28	$154,080	$0.28
Labor	14,400	0.42	43,200	.08
Water/elect./gas	2,160	0.06	30,240	.06
Enzymes/yeast	1,890	0.06	24,490	.05
Maintenance/repair	180	0.01	2,880	.01
Total	$28,260	$0.83	$254,880	$0.48
Fixed/variable cost	$45,339	$1.34	$281,853	$0.53
Less by-product credit	−1,980	−.06	−31,680	−.06
Total cost	$43,359	$1.28	$250,173	$0.47

Source: Bowker and Griffin (1983).
Note: Assuming a 10% discount rate and 12-year planning horizon

be 50 percent more than for the smaller plant. On a per-liter basis, insurance and property taxes account for approximately 10 percent of fixed costs in either case. As further evidence of economies of size, the per-liter fixed cost of insurance and property taxes for the smaller plant is in excess of ten times that for the larger (Bowker and Griffin, 1983).

The substrate cost per liter of ethanol for either plant is the same, whatever feedstock is used; there appear to be few opportunities for advantages of quantity purchases by the larger plant with respect to the substrate. Among grains, technical data and 1981 average prices indicate that sorghum would be a better choice than corn. Grain sorghum leads to a lower (by $0.02) per-liter ethanol production cost than corn.

Similarly, the by-product credit per liter of production is estimated to be the same for both operations. It should be noted, however, that for the smaller plant, the by-product credit reduces average total cost by only 4 percent, whereas the by-product credit for the larger plant reduces average total costs by as much as 11 percent (Bowker and Griffin, 1983).

The per-liter ethanol cost for the 34,000-liter-per-year plant in this study amounts to approximately three times that for the 545,000-liter-per-year plant. This evidence suggests that at present it is not economically rational to build and operate ethanol plants in the smaller range. This is particularly true in cases where operation is limited to one shift per working day (Bowker and Griffin, 1983).

Sorghum Residue Feedstock

This section evaluates ethanol production from grain sorghum residue. Table 7.16 shows the cost to bale grain sorghum residue and haul it various distances to the plant (Purdue University, 1979). Both an average residue of 5.3 tonne/ha (4,750 lb/ac) and a high-yield residue of 9.0 tonne/ha (8,000 lb/ac) were considered.

The cost per tonne of grain sorghum residue to an ethanol plant, not including nutrient loss or any payment to farmers for the residue, suggests that the haul distance will need to be relatively short. This analysis is based on a 24.1-kilometer (15-mile) haul distance for the average-residue yield and 16.1 kilometers (10 miles) for the high-residue yield.

Table 7.17 presents some characteristics of a large-scale ethanol plant using grain sorghum residues. Again, average- and high-yield residue production values are included. With 9.0 tonnes of residue per ha, 1,470 liters of ethanol could be produced per ha (157 gal/ac).

TABLE 7.16 Energy Use and Cost to Bale and Transport Grain Sorghum Residue to Ethanol Plant

	5.3 Tonne Yield/ha		9.0 Tonne Yield/ha	
Distance (km)	Diesel (L per tonne)	Cost ($ per tonne)	Diesel (L per tonne)	Cost ($ per tonne)
8.0	7.9	13.99	6.7	11.79
16.1	9.1	15.16	7.9	12.96
24.1	10.5	16.32	9.1	14.12
32.2	11.7	17.49	10.0	15.24
40.2	12.9	18.66	11.2	16.40
48.3	14.2	19.82	12.5	17.37
56.3	15.4	20.99	13.8	18.73
64.4	16.6	22.15	15.0	19.90
72.4	17.9	23.32	16.3	21.07
80.5	19.1	24.49	17.5	22.23

Source: Lacewell et al. (1982). Based on cost equations in Purdue University (1979).
Note: Labor rate calculated at $5/hour, fuel cost at $0.26/L ($1/gal).

A plant producing 75.7 Ml (20 million gal) per year would require residue from 51,500 hectares of land yielding 9.0 tonne/ha, as compared to 87,000 hectares at the average yield level. The cost of ethanol production in a large plant is estimated to be $0.40/L ($1.52/gal) with a residue yield of 9.0 tonne/ha, and $0.42/L ($1.59/gal) with a residue yield of 5.3 tonne/ha. There is no allowance for the cost of storage or of residue handling at the plant, and a zero price is placed on the residue itself. On the other hand, no credit is given for the value of by-products from the process, such as cattle feed (Lacewell et al., 1982).

Cellulose Feedstock

With new technology that would reduce conversion costs, ethanol production from cellulose appears to be economically attractive. At this time, however, the process offers great potential but is not at a point of commercialization.

Plant Oils as Diesel Fuel Substitutes

Plant oils are renewable resources and have the potential for direct use as a diesel fuel substitute without requiring extensive engine

TABLE 7.17 Characteristics of Large-Scale Ethanol Plant Using Grain Sorghum Residue

Item	Residue Yield per Hectare	
	5.3 Tonnes	9.0 Tonnes
Hectares of residue	87,000	51,500
Ethanol (L/ha)	870	1,470
Ethanol (GJ/ha)	14.5	24.6
Average residue haul (km)	24	16
Bale and haul ($/tonne)	15.93	12.96
Bale and haul (L diesel/L ethanol)	0.06	0.05
Conversion ($/L)[(a)]	0.32	0.32
Bale and haul ($/L)	0.10	0.08
Cost, not including residue ($/L)[(b)]	0.42	0.40

Source: Lacewell et al. (1982).
Note: Assumes a plant producing 20 million gallons (75.7-ML) per year, where 6 kg of residue produce one L of ethanol (Purdue University).
(a) Purdue University (1979).
(b) Each one dollar per tonne of residue required by the farmer increases ethanol costs about $0.01/L. This cost does not include any allowance for storage and handling of residues at the plant.

adjustment and modifications. The purpose of this section is to compare the past and current prices of diesel fuel and various plant oils on an equal basis, and then examine the national implications of substituting plant oils for diesel fuel.

Price Comparison

In general, plant oils contain approximately 94 percent of the energy in diesel fuel per unit of volume. Thus, it is not appropriate to simply compare prices on a per-liter basis. Table 7.18 calculates the price per 139 MJ for four plant oils diesel fuel, thus comparing prices on the basis of energy equivalence. Based on prices reported in the *Wall Street Journal* for February 1, 1983, the diesel-equivalent cost of the plant oils ranges from $1.59 to $2.41 per 139 MJ. It should be noted that the quoted prices were for bulk quantities at specific locations and exclude the additional transportation charges and profit margins required to retail these products. They also exclude taxes. On the

TABLE 7.18 Cost per Unit for Selected Plant Oils and Diesel Fuel

Item	Price per 139 MJ
Corn oil	2.35
Cottonseed oil	1.65
Peanut oil	2.41
Soybean oil	1.59
Diesel fuel (No. 2)	.79

Source: *Wall Street Journal*, February 1, 1983.

same basis, the current value of No. 2 diesel is approximately \$0.79 per 139 MJ. Therefore, it is clear that plant oils are about two to three times as expensive as diesel fuel at the present time.

Engler et al. (1982) present an identical analysis for monthly prices for the period 1970 to 1980; the results are replicated in Figure 7.1. The lower curve represents the historical trend of diesel prices during the period. The upper curve depicts the four plant oil prices—corn oil, cottonseed oil, peanut oil, and soybean oil—averaged to obtain a single curve. All prices are expressed for 139 MJ. It is evident from this graph that plant oil prices have been substantially above diesel prices over the entire period. In absolute terms, the amount of the discrepancy does not appear to have decreased much during this period. In terms of percentage differences, diesel fuel prices are gaining ground on plant oil prices: in the early 1970s, the ratio of the price of plant oils to diesel fuel was about 10:1; it was near 2:1 in 1981. Nevertheless, these graphical devices clearly demonstrate that plant oils have not been inexpensive substitutes for diesel fuel (Engler et al., 1982).

Sunflower oil prices are not reported daily or monthly, so sunflower oil has been excluded from consideration. However, annual sunflower oil prices are available for the past few years. Converting these prices to diesel equivalents results in values that lie within the other plant oil prices shown in Figure 7.1. Therefore, the previous results appear to apply equally to sunflower oil.

Given the recently published results of other researchers, the small-scale processing of sunflower seed does not appear to offer a more promising alternative (Hoffman et al., 1981). After examining the inputs required to operate over a 1,000-hour period, a press cap-

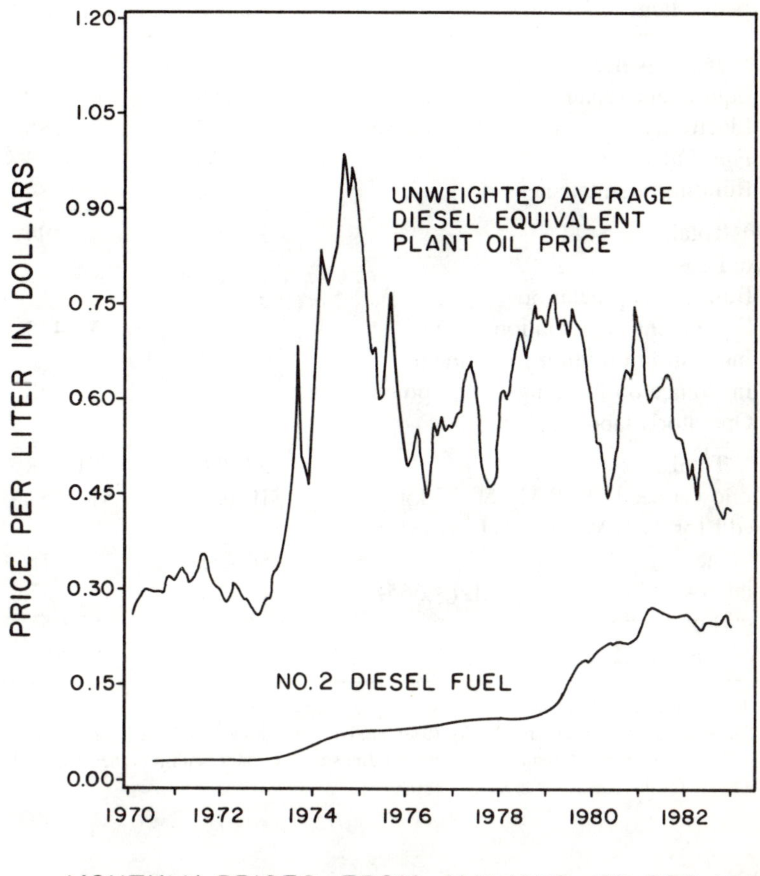

FIG. 7.1. Price comparison between major plant oils and diesel fuel (from Engler et al., 1982).

able of producing 13.9 L (3.6 gal) of sunflower oil per hour, these researchers reported a production cost of \$0.58/L (\$2.20/gal) if the costs of labor and the cost of the building are excluded. If labor and building expenses are included, the cost increases to \$1.06/L (\$4.01/gal). These results are given in Table 7.19. Neither estimate has been adjusted to a

TABLE 7.19 Estimated Cost of Processing Sunflower Oil Using a Small Press[(a)]

Expense Item	Scenario 1	Scenario 2
Variable expense		
Equipment repair	$ 780	$ 780
Electricity	185	185
Fuel Oil	345	345
Building maintenance	184	184
Total	$1,494	$1,494
Fixed cost		
Building depreciation	$ 369	
Equipment depreciation	415	$ 415
Interest on building investment	461	
Insurance on building & equipment	688	688
Operator's labor: 1,158.5 hr	5,792	
Total	$7,780	$1,128
Sunflower seed, 43.8 Mg $0.23/kg	$10,051	$10,051
Credit for 31.0 Mg of meal ($148/tonne)	$4,618	$4,618
Total cost to process 13.9 kL (3,665 gal) of sunflower oil	$14,707	$8,055
Per-liter cost of sunflower oil	$1.06	$0.581

Source: Hoffman et al. (1981).
Notes: The small press will crush 42.6 kg (94 pounds) of sunflower seed per hour. Scenario 1 includes all costs; scenario 2 assumes that the buildings are available with no other use, and no charge is made for the operator's labor.

diesel-equivalent basis. This adjustment inflates the costs to $0.63 and $1.51/L ($2.39 and $4.36/gal), respectively. Utilizing labor valued at $5 per hour resulted in a labor cost of $0.42/L ($1.58/gal) of oil. According to this information, labor is the greatest single expense in the crushing operation (Engler et al., 1982).

Implications of the Substitution

If substantial quantities of plant oils are diverted from traditional markets for use as fuel, the market impacts can be expected to be large and wide ranging. This section summarizes a recent study by Collins,

Griffin, and Lacewell (1982) on the impact of substituting cottonseed oil and soybean oil for diesel fuel. To examine the impact, a regional field crop and national livestock econometric model (TECHSIM) was used (Collins, 1980). The regions represented by TECHSIM are shown in Figure 7.2. Two scenarios, representing a 5 and a 10 percent replacement of agriculture's diesel fuel use by plant oils, were simulated for the years 1982–90. The base simulation assumes that no plant oils will be diverted to replace diesel fuel, whereas each policy simulation reflects a program position of purchasing plant oils annually to meet targeted diesel fuel shortage. The differences between the base simulation and each of the two policy simulations provide estimates of multimarket impact (Collins et al., 1982). The estimated impact of substituting plant oils for diesel fuel on the price of crops, farmer profit, and consumer surplus, as well as the regional effect of the planting shift, are discussed below.

Price of Crops. Purchasing activities for plant oils resulted in increased prices of cottonseed and soybean oils for both policy scenarios, which resulted in price increases for cottonseed and soybeans (Table 7.20). These price increases provided incentives to increase planted hectarage of plant oil crops. However, since meal is a joint product with oil, the prices of cottonseed meal and soybean meal decreased. Similarly, since cottonseed and cotton lint are joint products, an increase in production of cotton to produce more cottonseed depressed the price of cotton lint (Collins et al., 1982).

The prices of corn, small grains, and grain sorghum also increased as a result of a shift of producers from these crops to cotton and soybeans. However, the price increases of these crops were relatively small under both policy scenarios (Table 7.20).

A decrease in the prices of cottonseed meal and soybean meal feedstuffs encouraged livestock producers to increase the number of animals fed, which led to higher production levels of such livestock as fed beef, pork, and sheep—and to lower prices. Only non-fed beef increased in price, which is a result of a shift from non-fed livestock production to fed livestock production (Collins et al., 1982).

Farmer Profit. The total regional and national net returns or rents to fixed-production factors for farmers are given in Table 7.21. All re-

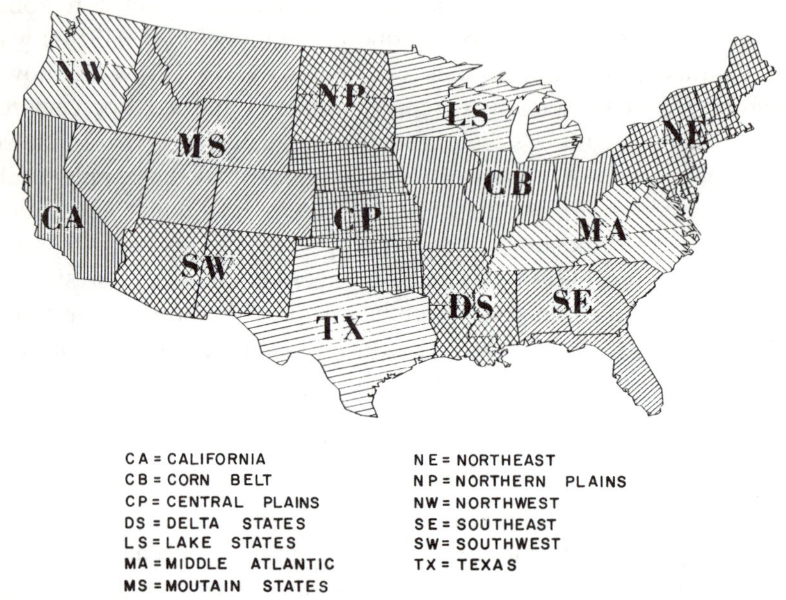

FIG. 7.2. Production regions within TECHSIM (from Collins, Griffin, and Lacewell, 1982).

gions experienced an increase in farmers' net returns (profit) under both program policies. The largest dollar increase was in the Corn Belt (CB) region, where producers gained \$615 and \$469 million in 1983 and 1990 with a 5 percent program policy, and \$1,283 and \$1,043 million in 1983 and 1990 under a 10 percent program policy. This region accounted for approximately one-third of the total increase in national net returns of farmers for both policies. In general, the Corn Belt (CB), Delta States (DS), Central Plains (CP), and Texas (TX) producers gained the most from diverting cottonseed and soybean oils to use as a diesel fuel substitute (Collins et al., 1982).

Consumer Effect (Welfare Shifts). The change in surplus measures for each diversion policy is illustrated in Table 7.22. The first measure depicts the change in total crop producer net returns or rents across all of the U.S. These are the rents resulting from summing the change in total regional field crop rents. Total crop rents beyond the farm gate

TABLE 7.20 Price Shifts of U.S. Field Crops and Livestock Associated with Diversion of Plant Oils to Diesel Use

		5% Diversion[a]		10% Diversion[a]	
Commodity	Unit	1982	1990	1982	1990
Corn	($/kg)	12.88	.16	.28	.20
Small grains	($/kg)	0.0016	0.0059	0.0024	0.0102
Grain sorghum	($/kg)	0.0004	0.0028	0.0008	0.0055
Cotton lint	($/kg)	−0.0132	−0.0430	−0.0573	−0.0683
Cottonseed	($/tonne)	66.41	67.45	192.74	193.44
Cottonseed meal	($/tonne)	−29.87	−59.09	−48.17	−86.09
Cottonseed oil	($/kg)	0.1658	0.1733	0.4808	0.4987
Soybeans	($/kg)	0.0453	0.0197	0.0787	0.0390
Soybean meal	($/tonne)	−39.57	−77.71	−63.60	−113.01
Soybean oil	($/kg)	0.4317	0.4350	0.7485	0.7623
Fed beef	($/kg)	−0.0029	−0.0004	−0.0049	−0.0007
Nonfed beef	($/kg)	0.0007	0.0026	0.0011	0.0035
Pork	($/kg)	−0.0196	−0.0216	−0.0320	−0.0302
Sheep	($/kg)	−0.0185	−0.0225	−0.0302	−0.0311

Source: Collins et al. (1982).

(a) The diversion level represents the amount of diesel fuel energy used in agriculture that is replaced by plant oils.

refer to net returns to intermediate industries that utilize field crops as inputs; they are not accounted for in other rent measures in Table 7.22. The total of meal and oil industry rents depicts the rent to meal and oil processors plus forward industry rents and final consumer surpluses that utilize oil and/or meal as an input. The other rent measures in Table 7.22 reflect the change in rents to other agricultural industries in TECHSIM. The total cost of purchasing plant oils reflects the purchase price times the quantity purchased for both cottonseed and soybean oils. The total welfare measure is the sum of all rent measures, less the cost of purchases for each policy. For a detailed explanation of how these rents were computed, see Chavas and Collins (1982).

As shown in Table 7.22, all producer/consumer classifications were net losers with the exception of field crop producers and final consumers of livestock products. The crop loss to industries beyond the gate resulted from more increases than decreases in crop prices. For example, cotton lint was the only field crop that experienced a

TABLE 7.21 Regional Shifts in Hectares and Net Returns: 1983 and 1990

Policy	Year	U.S.	Regions[a]												
			NW	CA	MS	SW	CP	NP	TX	LS	CB	DS	SE	MA	NE
		5% Diversion													
Hectare (1,000)															
Corn	1983	−429.4	—[b]	−1.2	−5.7	—[b]	−14.2	−16.2	−8.5	−11.3	−363.8	−0.4	−2.8	−2.4	−2.4
	1990	−436.7	—[b]	−0.8	−8.5	—[b]	−76.1	−8.9	−10.1	−40.5	−281.7	−1.6	−2.8	−2.8	−2.0
Small grains	1983	−10.9	2.4	0.8	12.1	−2.4	−38.9	—[b]	−17.4	14.6	11.7	0.8	1.2	2.0	2.0
	1990	61.5	10.1	2.0	32.4	−1.2	−91.1	−0.4	13.4	31.6	64.8	3.6	5.7	8.5	8.9
Grain sorghum	1983	14.6	0.0	−0.4	0.0	−8.5	9.7	—[b]	−17.4	0.0	0.4	−0.4	0.8	1.2	0.0
	1990	24.3	0.0	−0.4	0.0	−6.9	17.8	−0.4	−36.0	0.0	17.0	−3.2	1.6	2.4	0.0
Cotton	1983	22.7	0.0	8.1	0.0	9.7	6.1	0.0	22.3	0.0	0.0	−17.0	−1.6	−49.0	0.0
	1990	32.8	0.0	6.5	0.0	69.0	4.5	0.0	15.8	0.0	0.0	−3.6	3.2	−0.4	0.0
Soybeans	1983	717.5	0.0	0.0	0.0	0.0	28.7	24.3	12.1	102.4	355.7	34.4	59.5	86.2	14.2
	1990	686.8	0.0	0.0	0.0	0.0	10.1	10.9	10.5	91.5	459.7	39.7	22.7	35.6	6.1
Net returns	1983	1809.0	23.0	78.0	44.0	41.0	201.0	80.0	153.0	153.0	615.0	217.0	79.0	101.0	24.0
($1,000,000)	1990	1420.0	28.0	65.0	51.0	33.0	149.0	87.0	119.0	92.0	469.0	196.0	56.0	66.0	9.0

		10% Diversion													
Hectare (1,000)															
Corn	1983	−746.3	—[b]	−2.8	−8.9	−0.4	−27.9	−27.9	19.4	−19.0	−627.3	−0.4	−4.5	−4.0	−3.6
	1990	−738.6	−0.4	−2.4	−11.7	—[b]	−126.7	−13.8	−19.0	−57.8	−495.8	−1.6	−3.6	−3.6	−5.0
Small grains	1983	−25.1	4.5	1.2	19.8	−7.7	−67.2	—[b]	−29.9	23.9	19.4	1.2	2.4	3.6	3.6
	1990	50.2	17.0	2.8	50.6	−4.5	−184.5	−0.8	−27.5	44.9	108.5	5.3	9.3	14.6	14.6
Grain sorghum	1983	−63.5	0.0	−0.4	0.0	−25.1	15.0	−0.4	−55.8	0.0	1.2	−0.8	1.2	1.6	0.0
	1990	13.8	0.0	−0.4	0.0	−19.4	24.7	−0.8	−25.9	0.0	32.8	−4.5	2.8	3.6	0.0
Cotton	1983	118.2	0.0	23.1	0.0	28.3	15.0	0.0	61.5	0.0	0.0	−15.0	8.5	−3.2	0.0
	1990	110.9	0.0	18.2	0.0	20.2	10.9	0.0	45.3	0.0	0.0	1.2	13.0	2.0	0.0
Soybeans	1983	1178.5	0.0	0.0	0.0	0.0	49.0	42.1	20.2	178.5	595.3	30.4	93.9	144.5	24.5
	1990	1159.4	0.0	0.0	0.0	0.0	21.9	21.0	21.4	180.9	795.6	0.4	39.3	66.8	12.1
Net returns	1983	3773.0	37.0	197.0	73.0	109.0	374.0	137.0	352.0	304.0	1283.0	477.0	171.0	209.0	50.0
($1,000,000)	1990	3113.0	47.0	170.0	89.0	89.0	289.0	152.0	296.0	215.0	1043.0	420.0	130.0	148.0	25.0

Source: Collins et al. (1982).

(a) Regional delineation is shown in Figure 7.2. For each region, net returns represent the sum of changes in net returns for all field crops. The national net returns figures represent the summation of all regions' net returns.

(b) Field crop acreage changes are less than 1,000 acres (404.7 hectares).

TABLE 7.22 Welfare Impacts of Diverting Plant Oils to Diesel Use

Rent or Cost Measure	5% Diversion[(a)] 1982 ($ million)	5% Diversion[(a)] 1990 ($ million)	10% Diversion[(a)] 1982 ($ million)	10% Diversion[(a)] 1990 ($ million)
Total crop producer rents	1,914	1,420	3,666	3,112
Total crop forward industry rents	−987	−788	−1,768	1,477
Total meal and oil industry rents	−1,297	−720	−2,565	−1,665
Total livestock producer rents	−63	−56	−104	−80
Total livestock wholesale and retail producer rents	−74	−65	−121	−92
Total livestock final consumer surpluses	140	124	230	173
Total program cost of purchasing plant oils	429	431	1,095	1,106
Total welfare: sum of all rents less program purchase cost	−796	−516	−1,757	−1,135

Source: Collins et al. (1982).
(a) The diversion level represents the amount of diesel fuel energy in agriculture that are replaced by plant oils.

price decrease. Furthermore, since field crops are basically intermediate products, higher crop prices imply higher input costs to the industries that process these products. Thus, in the absence of higher forward products prices for those industries that use field crops as inputs, rents to these industries fell.

Although not shown, the meal and oil processors' proportion of the total meal and oil industry rent was positive. However, the increased prices of cottonseed and soybean oils also resulted in higher input costs to the industries that utilize these oils as inputs. Consequently, the total meal and oil industry rent was negative, even though it was offset somewhat by positive rent increases in the cottonseed and soybean meal markets.

Total livestock producer rent was negative, although some feedstuffs decreased in price and livestock production increased. Thus, the proportional decrease in livestock prices was greater than the lower feed input costs, resulting in rent losses. In addition to crop producers, final consumers of livestock products showed a positive increase in welfare due to lower livestock prices.

The program cost of purchasing plant oils ranged from approximately one-half billion dollars per year with a 5 percent diversion to over a billion dollars per year under a 10 percent diversion policy. This study does not take into account the program revenue that would result from selling plant oils as diesel fuel, but even if the program broke even (with the value of sales equaling the cost of the procurement and processing of plant oils for use as fuel), the total economic impact would be negative under both policy alternatives (Collins et al., 1982).

Regional Effect. Regional planting shifts due to the two policy alternatives for the years 1983 and 1990 are shown in Table 7.21. All results corresponding to 1983 reflect the initial impact of diverting plant oils, whereas the 1990 figures depict the final simulated impact. As expected, hectares of cotton and soybeans increased as a result of purchases of cottonseed and soybean oils. The largest single shift of planted hectares for both diversion policies was in the Corn Belt (CB) region, where producers shifted from corn production to primarily soybeans. Initially, U.S. field crop producers planted more hectares for all field crops except corn, small grains, and grain sorghum; however, only corn hectares decreased across all years for both plant oil diversion programs. Summing the changes in national planted hectares shows that approximately 285,300 and 368,700 more were planted in 1983 and 1990 with a 5 percent plant oil diversion policy, and approximately 461,700 and 595,700 more hectares with a 10 percent diversion policy. Although these appear to be large acreage shifts, they represent only an additional 3 to 6 percent increase in planted hectares for the U.S. (Collins et al., 1982).

Summary and Conclusions

This chapter is a summary of results from economic analyses of low-energy gas, ethanol, and plant oil production from agricultural biomass (mainly in the form of agricultural residues). The economic potential and some of the economic implications of agricultural biomass as an alternative energy source are assessed. In general, results indicate that under current conditions the only economically feasible alternative energy sources are from agricultural residues that are already harvested and collected: for example, cotton gin trash. It would appear that traditional food and feedstuffs (corn, sorghum, soybeans, and so forth) are costly energy feedstocks, while agricultural byproducts and wastes (gin trash, manure, stalks) are economically more promising. Moreover, concentrated wastes (gin trash, manures) save on transportation costs and often offer disposal benefits. Large commercial facilities are universally more economic than on-farm small-scale units, and the comparative advantage of large operations will likely prevent small-scale production of energy for sale from becoming economically feasible.

The results for the irrigation farm analysis indicate that energy production potential from residues and gin trash is sufficient to offset total energy use on the farm where improved management techniques are being applied. However, gasification and electricity generation are more costly than with conventional fuels and thus are not economically attractive at this time. In addition, there are problems associated with the storage of materials and the distribution of gas and electricity. As fuel prices adjust and new technology is developed, some of the on-farm energy production processes may become more economically competitive in the future.

Based on a very large gin of 40,000 bales (9,072 tonnes) per year, the on-site use of gin trash to produce electricity results in an estimated cost of $0.053 per kWh. However, about 70 percent of the electricity must be sold to the grid for a buy-back price ($0.02 kWh was used in this study), leaving more than $100,000 of costs not covered.

For a hog farm with a readily available supply of natural gas or propane, the anaerobic digester for gas production is not currently economically feasible. The net present value of gas production for replacing propane is negative in all cases. However, the anaerobic digester as a replacement for electricity is economically feasible in one scenario.

Two alternative plant sizes were considered, producing, respectively, 34,000 L (9,000 gallons) and 545,000 L (144,000 gallons) per year of ethanol from grains. The average production cost per liter of 182-proof ethanol was $1.40 ($5.28/gal) for the 34,000-L/yr plant and $0.47/L ($1.77/gal) for the 545,000-L/yr plant. Production costs per liter were reduced by about 67 percent in the larger plant. Ninety percent of the difference between per-liter cost for the two plants was accounted for by labor and plant utilization: the increased capacity of the larger plant, and its continuous operation. The results for ethanol production costs are consistent with those of other recent studies: under present conditions, 180- to 190-proof ethanol is unlikely to be produced by an on-farm plant when all costs are taken into consideration.

Production cost per liter of ethanol in a large plant using grain sorghum residue was $0.23/L for a residue yield of 9 tonne/ha, and $0.25/L with a residue yield of 5.3 tonnes/ha. Thus, the process of converting grain sorghum residues to ethanol offers great potential, but there are serious questions relative to the technology of conversion, storage, management, and handling of residues over several months.

The recent interest in developing plant oils as a substitute for diesel fuel is not expected to be implemented in the private sector because, given current market prices, the use of plant oils in diesel engines would cost two to three times as much as diesel fuel on a comparable energy basis. The record of prices from 1970 through 1980 indicates that this situation has existed for a long time; plant oils have never been inexpensive substitutes for diesel fuel. The relative price of diesel fuel would have to increase substantially before plant oils could become competitive on an economic basis.

The results of this study indicate that the impacts on domestic markets of substituting plant oils for 5 and for 10 percent of diesel fuel needs in agriculture would be large and generally disadvantageous. Such a program would bring about large price and profit shifts throughout the agricultural sectors. Field crop producers and oilseed processors would stand to gain, but most other agricultural industry profits would be reduced. Thus, it appears that a government-sponsored plant oil diversion policy would be costly and possibly inequitable.

The economic analysis of energy from biomass was primarily based on published material. For any system to compete economically, it must operate several months of the year. This means that energy (gas or electricity) will sometime be produced when it is not needed on the farm. The question of the cleanup of gas produced from residues or animal wastes was not addressed. Further, it was assumed that the gas and electricity could be sold on the wholesale market without difficulty or cost. Thus, before any of the energy generation systems is implemented, a more detailed study is needed. A careful analysis of cash flow, labor requirements, and maintenance costs must precede an investment analysis.

These analyses do not include all possible energy-generating systems; in addition, they are restricted by (1) assumed prices (which can rapidly change, and (2) the assumption that each system will operate as outlined (although operational problems exist for many of the scenarios). Moreover, they do not consider any risk or significant shortfalls in traditional fuels. Thus, although most systems considered are not economically feasible at present, many indicate much promise. Continued research and development are critical.

References

Atwood, J., and L. Fischer (1980). Cost of production of fuel ethanol in farm size plants. Report No. 115. Lincoln, Neb.: Department of Agricultural Economics, University of Nebraska.

Beck, S. R., and H. W. Parker (1979). Assessment of energy from biological processes. Draft report to Office of Technology Assessment, United States Congress, Washington, D.C.: OTA.

Bowker, M., and R. C. Griffin (1983). *The economic feasibility of small-scale ethanol production in Texas*. CEMR MS-6. College Station, Tex.: Center for Energy and Mineral Resources, Texas A&M University.

Butler Research and Engineering Co. (1980). Alcohol fuels workshop. Materials prepared for workshop sponsored by Farmers Home Administration, USDA and Economic Development Administration, U.S. Department of Commerce, Washington, D.C.

Chavas, J. P., and G. S. Collins (1982). Welfare measures from technological distortions in general equilibrium. *Southern Economic Journal* 48 (3): 745–53.

Collins, G. S. (1980). An econometric simulation model for evaluating aggregate economic impacts of technological change on major field crops. Ph.D. dissertation, Texas A&M University, College Station, Tex.

Collins, G. S., R. C. Griffin, and R. D. Lacewell (1982). *Impacts of substituting plant oils for diesel fuel*. Texas Agricultural Experiment Station Technical Article 17889. College Station, Tex.: Texas A&M University. (Preliminary results were presented in an earlier paper: National economic implications of substituting plant oils for diesel fuel. In *Proceedings, International Conference on Plant and Vegetable Oils as Fuels*. Fargo, N.D.: American Society of Agricultural Engineers, 1982.)

Dobbs, T., R. Hoffman, and A. Lundeen (1981). *Framework for examining the economic feasibility of small-scale alcohol plants*. Economics staff paper, Series No. 81-3. Brookings, S.D.: Economics Department, South Dakota State University.

Engler, C. R., W. A. LePori, L. A. Johnson, R. C. Griffin, K. C. Diehl, D. S. Moore, R. D. Lacewell, C. G. Coble, E. W. Lusas, and E. A. Hiler (1982). *Economic and engineering evaluation of plant oils as a diesel fuel*. Final report prepared for Texas Energy and Natural Resources Advisory Council. College Station, Tex.: Texas Agricultural Experiment Station, Texas A&M University.

Farmer Cooperatives (1981). Size makes operation of fuel alcohol still on FS farm impractical; continue study.

Federal Land Bank (1980). Federal Land Bank borrower characteristics. Unpublished monograph.

Fischer, L. K. (1980). The economics of producing fuel alcohol in farm-size plants. In *Proceedings of Alcohol Fuel Symposium*, 26–39. College Station, Tex.: Texas A&M University.

Hoffman, V., K. Kaufman, D. Helgeson, and W. E. Dinusson (1981). *Sunflower for power*. Fargo, N.D.: Cooperative Extension Service, North Dakota State University.

Lacewell, R. D., C. R. Taylor, and E. A. Hiler (1981). *Energy generation from cotton gin trash: An economic analysis*. CEMR Monograph Series. College Station, Tex.: Center for Energy and Mineral Resources, Texas A&M University.

Lacewell, R. D., E. A. Hiler, S. M. Masud, and R. D. Kay (1982). *Assessment of integrated agricultural and energy production systems*. CEMR Monograph Series. College Station, Tex.: Center for Energy and Mineral Resources, Texas A&M University.

Lacewell, R. D., and J. B. Penson, Jr. (1982). *Economic analysis of an anaerobic digester based on the Del Valle hog farm system concepts*. Report to Texas Energy and Natural Resources Advisory Council and SUMX Corporation. College Station, Tex.: Department of Agricultural Economics, Texas A&M University.

Masud, S. M., R. D. Lacewell, and E. A. Hiler (1983). Economic implications of cotton gin trash and sorghum residues as alternative energy sources. *Energy in Agriculture* 1 (3): 267–80.

Miller, F. L. (1980), Department of Soil and Crop Science, Texas A&M University, College Station, Tex. Personal communication.

Moore, D. S., R. D. Lacewell, and C. B. Parnell (1982). *Economic implications of pelleting cotton gin trash as an alternative energy source*. TAES B-1382. College Station, Tex.: Texas Agricultural Experiment Station, Texas A&M University.

National Climatic Center. Climatological Data, Texas. Environmental Data and Information Service, National Oceanic and Atmospheric Administration, U.S. Department of Commerce, Asheville, N.C.

Nichols, T. E., and J. D. Jackson (1981). *Economics of small-scale on-farm alcohol distilleries*. Economic Information Report No. 64. Raleigh, N.C.: Department of Economics, North Carolina State University.

Office of Technology Assessment (1980a). Agriculture, unconventional crops, and select biomass wastes. In *Energy from biological processes*, vol. 3, Appendices, Part B. Prepared by faculty and staff of the School of Agriculture, Purdue University. Washington, D.C.: OTA.

Office of Technology Assessment (1980b). Technical and environmental analysis. In *Energy from biological processes*, vol. 2. Washington, D.C.: OTA.

Oursbourn, C. D., W. A. LePori, R. D. Lacewell, K. Y. Lam, and O. B. Schacht (1978). *Energy potential of Texas crops and agricultural residues*. TAES MP-1361. College Station, Tex.: Texas Agricultural Experiment Station, Texas A&M University.

Pimentel, D., et al. (1983). Biomass energy review (based on report of DOE Energy Research Advisory Board, Biomass Panel). *Solar Energy* 30 (1): 1–31.

Pollard, K. (1982). Department of Agricultural Engineering, Texas A&M University. Personal communication.

Purdue University (1979). The potential of producing energy from agriculture. Draft report to Office of Technology Assessment, United States Congress, Washington, D.C.: OTA.

Smith, K. D., J. Philbin, L. Kulik, and D. Inman (1979). *Energy from agriculture*. Draft report to Office of Technology Assessment, United States Congress. Washington, D.C.: OTA.

SUMX Corporation (1982). Estimated cost for construction and operation of various sized anaerobic digestion facilities based on the Del Valle hog farm system concepts. Prepared by James R. Kidwell and David A. Malish, Austin, Tex.

CHAPTER 8

Conclusions

BILL A. STOUT and EDWARD A. HILER

Summary

A high standard of living and a vigorous economy depend on ample energy supplies at reasonable prices. No single fuel source is likely to meet future energy needs. A diversity of options is desirable.

Since fossil fuel supplies are finite, continued human existence and well-being will eventually depend on renewable or inexhaustible energy forms such as solar energy. Biomass—produced from photosynthesis—is one form of stored solar energy. The millions of tons of biomass potentially available for fuel could provide up to 5 EJ of energy annually (6 or 7 percent of the nation's energy needs).

Many processes have been developed for converting biomass to more useful energy or fuel forms. These conversion processes are described in detail in Chapters 2 through 6, and an economic analysis of biomass as an alternative energy source is presented in Chapter 7.

Thermochemical Conversion for Energy and Fuel

Thermochemical conversion processes include direct combustion, gasification, and pyrolysis. Direct combustion is an oxidation process involving excess air. The primary purpose of direct combustion is to release heat. Gasification is a partial combustion process, with restricted oxygen supply resulting in the production of combustible gases. Pyrolysis is a heating process conducted in the absence of oxygen. It may produce liquid, gaseous, or solid fuels.

Characterization of the feedstock is an essential prerequisite for scientific design of thermochemical conversion systems. Fuels are characterized by proximate analysis (percent moisture, ash, volatile matter, and fixed carbon) and by ultimate analysis (percent of each chemical element). Numerous biomass feedstocks are characterized in Tables 2.1 and 2.2.

Direct Combustion. Three factors are required for complete combustion of a fuel: (1) heat to initiate the reaction, (2) mixing of oxygen and air with fuel, and (3) sufficient time for completion of the reactions.

The theoretical quantity of air necessary to provide sufficient oxygen to combine completely with a given quantity of fuel is termed *stoichiometric air*. Air supply in addition to this quantity is termed *excess air*.

Slag formation is a serious problem in the combustion of certain biomass fuels (such as cotton gin trash) in excess air.

Direct combustion systems are generally classified as pile burners, semi-pile burners, suspension burners, and fluidized-bed combustors. Fluidized-bed technology was used for direct combustion and gasification research at Texas A&M University. A fluidized bed consists of a refractory-lined chamber with an air distributor at the base. Inert particles are contained in the chamber. Air is forced through the distributor plate with a velocity sufficient to suspend the inert particles. The suspended particles circulate within the chamber, forming a turbulent mass. When heated, the fluidized-bed particles quickly transfer heat to the fuel particles as they are introduced into the bed.

Because of slagging and fouling problems when burning high-ash-content biomass fuels, emphasis on direct-combustion research at TAMU was reduced in favor of gasification research.

Gasification. Biomass may be converted to combustible gases through partial combustion, with oxygen kept below the stoichiometric level. The useful gaseous products—carbon monoxide, hydrogen, and a small amount of methane—are referred to collectively as *producer gas*, a low-energy gas containing 10 to 20 percent of the heat content of natural gas.

Gasifiers are classified as moving-bed, fluidized-bed, rotary-kiln, and multiple-hearth types. The fluidized-bed gasifier used for research at TAMU is physically similar to the fluidized-bed combustion system described previously, but air and fuel are metered simultaneously to provide an air-to-fuel ratio of 30 to 40 percent of the theoretical requirement.

The TAMU gasifier/boiler consists of a 61-cm-diameter fluidized-bed unit with auxiliary equipment for controlling fuel feed and particulate removal, a gas-mixing chamber and burner, and a fire-tube boiler.

Tables and figures are presented in Chapter 2 to document operating conditions, gas output, and particulate emissions.

Pyrolysis. Pyrolysis is a thermochemical process, conducted in the absence of oxygen and at temperatures generally below 600°C, through which biomass is converted into liquid, gaseous, or solid fuel products. Reaction parameters such as heating rate, final temperature, residence time at temperature, particle size, moisture and ash content, and presence or absence of air or oxygen determine the yield and quality of the final products. Liquid tars or solid chars are usually the primary recovered products.

Product formation mechanisms are discussed in Chapter 2 along with process design guidelines. An idealized process flow sheet is presented, as well as gas chromatograms of the volatile constituents of tars derived from the updraft gasification of a number of agricultural residues. Subsequent hydrotreating of the tars may be used to produce similar products from dissimilar biomass feedstocks.

Biological Conversion and Fuel Utilization:
Anaerobic Digestion for Methane Production

Anaerobic digestion of livestock manure and other feedstocks for methane production is discussed in Chapter 3. Bacterial degradation of organic matter under anaerobic conditions releases a mixture of gases, which usually consists of 50 to 60 percent methane, 40 to 50 percent carbon dioxide, and numerous trace gases such as hydrogen sulfide. Biogas is considered a medium-energy gas, containing 19 to 22 MJ of energy per cubic meter.

Environmental Factors. Anaerobic digestion is a complicated biological process and therefore sensitive to a number of environmental factors such as temperature, pH, organic loading rates, influent-solids concentration, nutrient availability, and toxic substances.

Temperature is an extremely important factor that controls microbial growth rates during the fermentation process. Mesophilic bacteria are most active in the range of 28 to 42°C, whereas thermophilic bacteria are most active at temperatures above 42°C. The optimum pH range is between 7.0 and 7.2. Organic loading rates of 1.6 to 3.2 kg of volatile solids per day per cubic meter of digester, and an influent-solids concentration of 6 to 12 percent are recommended. The toxic substances of principal concern are heavy metals, high concentrations of alkaline metals (Mg, Ca, Na, K), ammonia, and soluble sulfides.

Several kinetic models of bacterial growth are presented. The biogas production system involves feedstock handling, materials preparation, the digester, gas-scrubbing equipment, gas storage, animal feed recovery, and fertilizer recovery.

Anaerobic digesters can be classified according to loading frequency, temperature, degree of mixing, configuration, and construction materials.

Operational modes include continuous-flow, batch-flow, plug-flow, and continuously expanding digesters. In the batch process, mixing may be either continuous or intermittent. In a plug-flow system, a longitudinal digester is used; influent is introduced at one end and effluent is withdrawn from the other end. No internal mixing is used in this process.

The performance of operating digesters is presented in tabular form in Chapter 3. Methane yield is a measure of both waste degradability and digester efficiency. It is the amount of gas produced per pound of organic solids converted to organic acids and biogas. On the average, about 0.44 to 0.50 m^3 of methane is produced for every kilogram of volatile solids actually destroyed. Methane yields usually range from 0.16 to 0.31 m^3 of CH_4 per kilogram of volatile solids added to the digester.

The methane production rate is the amount of gas produced per unit of digester volume. For mesophilic digesters, it ranges from about 0.03 to 0.12 m^3/day of methane per cubic meter of digester volume, and 3.0 to 3.8 $m^3/day/m^3$ for thermophilic digesters.

Gas-scrubbing systems may be needed to remove water vapor, hydrogen sulfide, or carbon dioxide. The end use of the biogas will determine the type of scrubbing necessary.

The most economical on-site uses of biogas are for electricity generation and boiler fuel. Methane can also be compressed and utilized as a vehicle fuel.

By-products of biogas production include slurry cake, which can be used as an animal feed, and digester slurry, which is used for fertilizer.

Biological Conversion and Fuel Utilization:
Fermentation to Produce Ethanol

Ethanol production by fermentation involves three general steps (Chapter 4). The first step is the conversion of starch into fermentable sugars by hydrolysis and saccharification. The second step is fermentation, during which yeast converts the fermentable sugars into ethanol. The final step is distillation (and dehydration) to separate the alcohol from water.

Starch Hydrolysis and Saccharification. Before fermentation can take place in starchy feedstocks, the carbohydrates must be broken down into fermentable sugars. Grains are normally ground in order to expose the starch molecules. Factors such as the water-to-grain ratio, temperature, pressure, pH, and holding period or residence time are important variables that affect starch conversion. Enzyme manufacturers provide recipes for optimum conditions for each enzyme. An acid hydrolysis process may be used in lieu of the enzymatic process.

Complex sugars are broken down into fermentable sugars by use of a second enzyme. Again, specific recipes must be followed for each enzyme.

Fermentation. After inversion has been accomplished, yeast is added to the mash to convert the simple sugars into ethanol; carbon dioxide is released as a by-product. Batch fermentation is traditional. The maintenance of appropriate temperatures and pH levels is essential. The beer produced in the fermentation process contains up to 12 percent ethanol.

Distillation and Dehydration. To be useful for fuel, the ethanol concentration must be increased to 80 to 100 percent (160 to 200 proof). Fractional distillation at atmospheric pressure is most commonly used for separating ethanol from water to produce concentration up to 95 percent.

Anhydrous or 100 percent (200-proof) ethanol can be obtained by several dehydration processes. The conventional method for obtaining an anhydrous product is tertiary azeotropic distillation, using a third component such as benzene. Further distillation of the solution containing benzene produces anhydrous ethanol.

Ethanol and water can also be separated by means of molecular sieves. The use of desiccants, such as calcium oxide or fermentable grains, has also been investigated.

Feedstocks. Any plant material that contains sugar or starch can be used as a feedstock for producing ethanol by fermentation. The most common feedstock candidates in the United States are grain, sweet sorghum, and sweet potatoes.

Research is underway on the conversion of lignocellulosic feedstocks, such as agricultural or wood residues. Plant cellulose is the most abundant material on earth. Development of cost-effective methods of hydrolysis of cellulose or hemicellulose feedstocks is a desirable yet elusive goal. Techniques for using acids and enzymes for the hydrolysis of cellulose and hemicellulose are the subject of intensive research.

Stillage Processing. The economic feasibility of ethanol fuel production from agricultural commodities depends heavily on the effective utilization of the nutrients from stillage. Stillage is the liquid residue from fermentation of grain or other feedstocks. It is recovered from the base of the beer-distillation column. While stillage from grain fermentation contains 5 to 10 percent solids.

The biochemical oxygen demand of stillage liquids is about two orders of magnitude higher than that of typical raw domestic sewage, so the pollution potential is high. The composition of stillage depends on the specific feedstock. Many examples are given in Chapter 4.

Ethanol Use. Ethanol has a high octane number and is generally suitable for fuel in spark-ignition engines. Anhydrous ethanol may be used as an octane booster by blending it with unleaded gasoline. Ethanol that contains more than a fraction of 1 percent water, referred to as wet ethanol, cannot be blended with gasoline without the danger of phase separation as the temperature fluctuates. Wet ethanol may be used without blending, following appropriate engine modifications.

Ethanol is not generally suitable as a fuel for diesel engines. However, research is underway to develop suitable fumigation techniques for introducing ethanol as a supplementary fuel in diesel engines, and facilitating the blending of ethanol with diesel fuel in emulsifiers.

Plant Oil Extraction and Utilization

Dr. Rudolph Diesel used plant oil as one of the fuels for the development of the compression ignition engine in the late 1800s. Numerous studies have been conducted over the years to evaluate plant oils as fuels for compression ignition engines. Chapter 5 examines oilseed processing technology, as it applies to fuel preparation, and presents engine test results.

Extraction Technology. Oil may be removed from oilseeds by two processes: screw pressing (expelling), and solvent extraction. Screw pressing is a simple process that can be applied on farm, but not all the oil is removed: up to 20 percent normally remains in the cake.

Solvent extraction is used in commercial plants with only 1 percent or less of the oil remaining in the meal. Hexane is the most common solvent used, but because of the hazardous nature of the chemical, the process is not recommended for on-farm operations.

Oil Processing. Crude plant oil contains impurities and has a high viscosity, rendering it unsatisfactory as diesel fuel. Processing operations include: (1) filtration to remove particulates that clog fuel lines and lead to engine deposits, (2) degumming to remove phosphatides and other gum materials, (3) alkali refining to remove or neutralize free fatty acids, (4) interesterification to lower the viscosity of plant oils to the viscosity range of petroleum-based diesel fuel, and (5) cracking to reduce oil-molecule size and produce a variety of lighter products.

Engine Testing. Short-term engine tests show that plant oils are satisfactory fuels for diesel engines in terms of power output and thermal efficiency. Sustained operation of engines with fuels containing more than 20 percent plant oil, however, frequently leads to injector fouling, ring sticking, polymerization of crankcase oil, and other operation problems. Plant oils are more suitable fuels for indirect-injection engines than for open-chamber or direct-injection engines. Most engine operating problems can be overcome by appropriate oil processing and selection of engine type or design.

Systems Engineering of Biomass Fuel Energy Alternatives

Many energy end uses require liquid or gaseous fuels. In order for biomass to be an effective alternative to fossil fuels, it must be converted into useful energy forms. These energy conversion processes must be reliable and environmentally safe. The entire biomass production, conversion, and utilization process must be economically competitive with other options.

Systems engineering is *engineering* directed at optimizing the system's performance. The phases of systems engineering include developing the system concept, preliminary design, prototype development, and designing the final system. Each phase incorporates feedback from preceding phases until the final system design is achieved.

One of the most promising biomass energy conversion alternatives is a small-power cogeneration system fueled with agricultural residues.

Simulation. Two examples are included in Chapter 6. The first is a small power plant located at a cottonseed oil mill. The electric power generated can be used to replace electric power purchased from a utility company. The second example uses cotton gin trash as a fossil fuel extender in a coal-fired power plant.

A current simulation study at TAMU involves a small power plant located at a cottonseed oil mill or a cotton gin. Flow diagrams, flow charts, and tables of input data are presented along with the projected cost of electricity under different operating conditions. The electricity costs may be as low as \$0.044/kWh when producing 20.3 GW·h/year with a 2.6-MW generator.

Another simulation study involves the use of cotton gin trash as a supplemental fuel for a large coal-fired power plant in west Texas. The model was composed of four subsystems: cotton harvesting, ginning, transportation, and the power plant. In this simulation, a total of 205,000 tonnes of gin trash from 231 cotton gins was delivered to the power plant and burned in 352 days. At \$4.54 per tonne of gin trash, the 231 gins in the study averaged a net loss of \$0.36 per tonne processed. At \$11.00 per tonne, these same gins averaged a net profit of \$3.96 per tonne processed. Ginners annually incur a net loss of \$6.50 to \$10.00 per tonne for conventional disposal. For the one million bales in this study (273,000 tonnes), use of gin waste in large coal-fired power plants could generate a savings of over \$2 million.

Energy-Integrated Farms. Integrated farm energy systems involve various alternative energy technologies (biogas from anaerobic digestion, ethanol, solar heating, and so forth) and energy conservation measures (waste heat recovery, low pressure irrigation, irrigation scheduling, conservation tillage, and so on). Energy-integrated farm systems use a combination of existing or developing technologies to reduce farm dependence on petroleum fuels.

Economic Implications of Agricultural Biomass as an Alternative Energy Source

The purpose of Chapter 7 is to evaluate the potential of agricultural biomass as an alternative source of energy and to assess the economic implications. The biomass options considered include (1) the use of on-farm cotton gin trash for production of low-energy gas; (2) on-farm and gin-site electricity generation from gin trash; (3) biogas production from hog manure through anaerobic digestion; (4) ethanol production from fermentation of corn grain, sorghum grain, and sorghum residues; and (5) the production of plant oils from sunflower, cottonseed, and soybeans for use as diesel fuel substitutes.

Gasification. The operation of a fluidized-bed gasifier of appropriate size was evaluated and input data presented for a farm, a large cooperative, and a cotton gin. The estimated cost of energy produced from an on-farm gasifier operating part time was $7.42/GJ as compared with the current natural gas price of $3.32/GJ. The cost of gas produced by a large farm cooperative was estimated at $3.58/GJ using a 907 tonne/day gin trash facility or a 2,720 tonne/day grain-sorghum residue facility.

An electric generator located at a cotton gin and using gin trash for fuel could produce electricity at a cost of about $0.053/kWh. However, since much of the electricity produced cannot be used on site and is sold to a grid at their cost of generation, the system incurs an annual loss of over $100,000.

Anaerobic Digestion to Produce Biogas. An economic analysis of the anaerobic digester system located at the Del Valle hog farm indicates a net present value of $2,743 over the ten-year evaluation period, if favorable financing is obtained and the tax liability kept low. With less favorable financing and increased tax liability, the results indicate negative net present values ranging from $3,546 to $9,126. The many assumptions that led to these values are presented in Chapter 7.

Ethanol Production. Economic analyses of ethanol plants indicate production costs of $0.47 to $1.27 per liter of 160- to 190-proof ethanol produced in plants with capacities in the range of 34,000 to 545,000 liters/year. A capital investment of $1.62 per liter of annual capacity was required for a 45,400 liter plant, $0.69 per liter for a 151,000 liter plant, and $0.32 per liter for a 250,000-liter plant.

Plant Oils as Diesel Fuel Substitutes. February, 1983, prices for plant oils ranged from $0.42 to $0.64/liter. The price for No. 2 diesel fuel was $0.21/liter. Thus, plant oils are currently two to three times more expensive than an equivalent amount of diesel fuel.

A national analysis of the economic implications of diverting plant oils to substitute for 5 and 10 percent of the diesel fuel use in agriculture suggests dramatic adjustments in cropping patterns, in returns to production agriculture and related industries, and in consumers' well-being.

The many assumptions and limitations leading to these economic conclusions are given in Chapter 7.

Generalizations

Agriculture has several goals, but its basic and fundamental purpose is to produce food so that everyone may have an adequate diet. Secondary purposes for agriculture are to provide employment and to produce fibers, structural materials, and biomass for energy. Proposals to produce biomass energy raise a number of issues, some old and others new.

A portion of the grain and forage produced in agriculture has traditionally been fed to the animals that provide draft power for tillage transport and other mechanical tasks. Thus, the concept that agriculture may produce a portion of its own energy needs is an old issue (Table 8.1). Another old issue is the question of the economic desirability of producing biomass fuels.

A number of new issues related to energy production from agriculture and forestry are also presented in Table 8.1 (Fluck, 1984).

Energy Balance

The first law of thermodynamics tells us that energy is neither created nor destroyed. If *all* forms of energy going into a process and *all*

TABLE 8.1 Considerations in Producing Energy from Agriculture

Old Issues	New Issues
Competition with food production	Energy balance
Economic desirability	Utilization of residues
	Environmental impact
	Magnitude of biomass energy production
	Scale of biomass conversion systems
	Diversity of energy sources
	Effect on overall economy

Source: Adapted from Fluck (1984).

forms of energy coming out are accounted for, the input and output will be equal; thus, a discussion of energy ratios or energy balance is meaningless.

Most agricultural energy analysts consider solar energy input—the energy that drives the hydrologic cycle to produce rainfall, and makes it possible for plants to produce biomass through photosynthesis—as energetically free; that is, solar energy is not calculated as an energy input. If this solar input is neglected and only the cultural energy input considered, plants produce energy gains on the order of 2:1 to 10:1. Thus, it is possible, though by no means assured, to produce biomass feedstock, convert it to more useful energy forms, distribute it to the point of end use, and still maintain an energy gain.

A net energy gain from biomass fuels, relative to petroleum input, is essential for a successful biomass fuel program. However, an overall net energy gain may not be necessary in the short run if a low-quality bulky fuel is upgraded to a high-quality, clean-burning fuel—especially an energy-dense liquid fuel to power existing mobile vehicles.

Utilization of Residues

Agricultural and forestry residues, including animal manures, may be burned directly to produce heat or may be converted to more useful liquid or gaseous forms by a number of conversion processes. Valuable by-products remain after certain conversion processes.

Environmental Impact

A considerable portion of organic matter must be returned to the soil to maintain soil tilth and carbon balance and to control erosion. The belowground portion of many common species of plants constitutes about half the total biomass and normally remains in the soil. The amount of aboveground residue that should be returned to the soil is a matter of discussion and controversy. Posselius and Stout (1983) developed a procedure for evaluating the organic matter requirements of individual fields based on the soil type, slope, field dimensions, and climatic conditions. The procedure determines the amount of organic matter that must be returned to the soil in order to limit water and wind erosion to acceptable levels.

Magnitude of Biomass Energy Production

The diversion of small quantities of agricultural or forestry production to biomass fuels will probably have a limited impact on existing markets for agricultural commodities and land requirements; however, large programs designed to provide a significant portion of the nation's energy from biomass may disrupt already established markets and cause undue pressure on available land, leading to degradation by wind or water erosion.

Scale of Biomass Conversion Plants

Small farm-scale biomass conversion plants are appealing because they would be under the direct control of the farmer, and the supply lines would be short. Fuel produced in farm-scale plants is relatively expensive, however, and often not competitive with current prices for petroleum fuels. Conversion processes are often not as simple as they appear, and may require knowledge and skills beyond those available on most farms.

Large industrial-scale biomass conversion plants can produce fuels at lower unit costs and can maintain better fuel quality control, but supply lines will be longer than for small plants. Transporting sometimes bulky biomass feedstocks over long distances adds to the cost. Pollution problems, too, may be magnified by increased centralization, but technological options for pollution control are also increased.

Diversity of Energy Sources

Excessive reliance on a few energy sources may be undesirable. A broad range of energy options is preferable. Biomass fuels add to the diversity of energy sources and thus reduce the impact of supply disruptions.

Impact on Overall Economy

Interactions between various sectors of the economy are complex and not always completely predictable. Biomass for fuel is a complex subject involving the growth, collection, densification, transport, conversion, and utilization of organic matter. Often, fuel production must compete with important alternative uses for biomass. The use of biomass for fuel will almost certainly have an impact on food, feed, and fiber prices. Biomass fuel programs will also affect land prices and may have other positive or negative impacts throughout the economy.

Concluding Remarks

Biomass energy involves a broad range of feedstocks including forest products, grains, oilseeds, residues, algae, and animal manure. A dozen or more conversion technologies may be classified as thermochemical, biological, or extraction processes.

A multitude of end uses involve heat production, internal combustion engine operation, and electrical generation. Readers who expect a simple, concise set of conclusions regarding the technical feasibility and economic viability of biomass energy systems will be disappointed. Such simplicity is both unwarranted and impossible because of the wide range of technologies being examined and the dynamic and uncertain nature of the economic system in which biomass is being considered as an energy source.

Fossil fuels (petroleum, natural gas, and coal) provide about 90 percent of the energy used in the United States today. It seems clear that if current prices prevail, biomass energy will not displace major quantities of fossil fuels in the next decade or two. In general, most biomass fuels are not economically viable under the current price structure.

There are, undoubtedly, many exceptions to the foregoing generalization. Forest products and residues are widely used as a source of industrial and residential heat, and several hundred million liters of fuel ethanol are sold each year at subsidized prices.[1] Feasibility studies for the gasification of cotton gin trash and other "collected" residues yield results that appear quite attractive.

Future energy prices and supplies are uncertain. Who could have predicted the tenfold increase in oil prices during the past decade? Energy prices are not established merely on the basis of supply and demand; unpredictable political decisions and military conflicts often overshadow predictable physical factors. If fossil fuel prices increase in the future, relative to other prices, then biomass fuels may become more attractive.

Shifts in the energy mix are not completely responsive to direct economic forces. Biomass fuels face a host of institutional and environmental barriers to their adoption that may take years to sort out, even if biomass fuels do become economically attractive. For example, no infrastructure exists for distributing and marketing biomass fuels (with the exception of ethanol-gasoline blends). Insurance coverage for biomass conversion plants and fuels is an uncertain area. And air and water pollution standards may limit both the production of biomass fuels and their utilization.

The possibility of energy shortfalls in the future casts biomass fuels in a somewhat different light. What can a farmer afford to pay for liquid fuels to plant and harvest crops if petroleum fuels are not available? If petroleum shortfalls occur and oil prices increase, the price ratio of petroleum and biomass products may shift, with biomass fuels becoming more attractive. Advocates of biomass fuels have pointed out the advantages of a diversified and locally controlled energy source. In Europe during World War II, engines were fueled with producer gas and alcohol when gasoline and diesel fuels were in short supply.

Research results, such as those reported in this monograph, constitute a tremendous wealth of technical and economic data. While economic analyses generally show that most biomass fuels are not economically viable today, the technical knowhow is available to produce and utilize biomass for fuel.

If a future energy shortfall occurs, or if relative prices shift to make renewable biomass energy alternatives economically attractive, a substantial amount of energy can be produced from biomass.

[1]Total U.S. fuel ethanol capacity, 872 million liters (*Fuel Alcohol USA*, Jan.–Feb. 1983, p. 13).

References

Fluck, R. C. (1984). Issues in energy farming. *Energy in agriculture*. Amsterdam: Elsevier Science Publishers.

Posselius, J. H., and B. A. Stout (1983). Crop residue availability for fuel. In *Liquid fuel systems*, ed. D. L. Wise. Boca Raton, Fla.: CRC Press.

The Authors

William H. Aldred is Associate Professor, Department of Agricultural Engineering, Texas A&M University

Charlie G. Coble is Professor, Department of Agricultural Engineering, Texas A&M University.

Glenn S. Collins is Visiting Assistant Professor, Department of Agricultural Economics, Texas A&M University

Richard P. Egg is Research Agricultural Engineer, Texas Agricultural Experiment Station

Cady R. Engler is Associate Professor, Department of Agricultural Engineering, Texas A&M University

David A. Givens is Research Assistant, Texas Agricultural Experiment Station

Ronald C. Griffin is Assistant Professor, Department of Agricultural Economics, Texas A&M University

Edward A. Hiler is Professor and Head, Department of Agricultural Engineering, Texas A&M University

Lawrence A. Johnson is Associate Research Chemist, Texas Engineering Experiment Station

Ronald D. Lacewell is Professor, Department of Agricultural Economics, Texas A&M University

Wayne A. LePori is Professor, Department of Agricultural Engineering, Texas A&M University

Sharif M. Masud is Visiting Assistant Professor, Department of Agricultural Economics, Texas A&M University

Calvin B. Parnell, Jr., is Professor, Department of Agricultural Engineering, Texas A&M University

Donald L. Reddell is Professor, Department of Agricultural Engineering, Texas A&M University

Ed J. Soltes is Professor, Department of Forest Science, Texas A&M University

Bill A. Stout is Professor, Department of Agricultural Engineering, Texas A&M University

JOHN M. SWEETEN is Professor, Department of Agricultural Engineering, Texas A&M University

C. MICHAEL YARBROUGH is Research Assistant, Texas Agricultural Experiment Station

Index